Writing Science

Pittsburgh Series in Composition, Literacy, and Culture
David Bartholomae and Jean Ferguson Carr, Editors

Academic Discourse and Critical Consciousness
Patricia Bizzell

Eating on the Street:
Teaching Literacy in a Multicultural Society
David Schaafsma

Fragments of Rationality:
Postmodernity and the Subject of Composition
Lester Faigley

The Insistence of the Letter:
Literacy Studies and Curriculum Theorizing
Bill Green, Editor

Knowledge, Culture and Power:
International Perspectives on Literacy as Policy and Practice
Peter Freebody and Anthony R. Welch, Editors

Literacy Online:
The Promise (and Peril) of Reading and Writing with Computers
Myron C. Tuman, Editor

PRE/TEXT: The First Decade
Victor J. Vitanza, Editor

Word Perfect: Literacy in the Computer Age
Myron C. Tuman

Writing Science: Literacy and Discursive Power
M.A.K. Halliday and J.R. Martin

Writing Science:
Literacy and Discursive Power

M.A.K. Halliday and J.R. Martin

University of Pittsburgh Press

UK The Falmer Press, 4 John St., London WC1N 2ET
USA University of Pittsburgh Press, 127 North Bellefield Avenue, Pittsburgh, Pa. 15260

The authors would like to acknowledge the financial support of the Australian International Literacy Year (ILY) Secretariat.

First published 1993 by The Falmer Press in the *Critical Perspectives on Literacy Series*, edited by Allan Luke.

Library of Congress Catalog Card Number 93-60340

Library of Congress Cataloging in Publication Data are available on request

ISBN 0-8229-1180-9 (cl)
ISBN 0-8229-6103-2 (pbk)

Jacket design by Benedict Evans
Typeset in 9.5/11pt Bembo
by Graphicraft Typesetters Ltd., Hong Kong

Printed in Great Britain by Burgess Science Press, Basingstoke on paper which has a specified pH value on final paper manufacture of not less than 7.5 and is therefore 'acid free'.

Contents

List of Tables and Figures   vii–ix

Introduction   x

Introduction: The Discursive Technology of Science   1

Chapter 1   General Orientation   2

Chapter 2   The Model   22

Part 1:   Professional Literacy: Construing Nature   51

Introduction   52

Chapter 3   On the Language of Physical Science   54

Chapter 4   Some Grammatical Problems in Scientific English   69

Chapter 5   The Construction of Knowledge and Value in the
             Grammar of Scientific Discourse: Charles Darwin's
             The Origin of the Species   86

Chapter 6   Language and the Order of Nature   106

Chapter 7   The Analysis of Scientific Texts in English and
             Chinese   124

Part 2:   School Literacy: Construing Knowledge   133

Introduction   134

Chapter 8   The Discourse of Geography: Ordering and Explaining
             the Experiential World   136

Chapter 9   Literacy in Science: Learning to Handle Text as
             Technology   166

Chapter 10   Technicality and Abstraction: Language for the
             Creation of Specialized Texts   203

Contents

Chapter 11   Life as a Noun: Arresting the Universe in Science and
             Humanities   221

References   268

Index   281

Tables

2.1 Solidarity across Example Systems, Metafunctions and
Contextual Categories   30

5.1 Some Grammatical Features in the Scientific Writings of
Chaucer and Newton   89

5.2 Common Types of Logical-Semantic Relation, with
Typical Realizations as Conjunction and Preposition   91

5.3 Examples of Lexicalization of Logical-Semantic Relations
(as Verbs)   91

5.4 Examples Showing Logical-Semantic Relations Lexicalized
(1) as Verb, (2) and (3) as Noun   92

5.5 Summary of Motifs Constructing Theme and New of
Ranking Clauses   101

8.1 Botanical Classification of One Variety of Rose   138

8.2 Taxonomic Organization of a Geography Textbook   147

8.3 Examples of Attributive Relational Process   154

9.1 How to Write Up Science Experiments   195

11.1 Four Basic Text Types in Science and Humanities   222

11.2 Congruent and Incongruent Realizations of Key Semantic
Variables   238

11.3 Ideational Meaning: Congruent and Incongruent
Realizations   239

11.4 Multiple Readings of Ideational Metaphor   239

11.5 Structures of Texts   258

11.6 Synoptic Overview of Key Meanings in the Pedagogic
Discourses of Science and History   266

Figures

2.1	Language as the Realization of Social Context	25
2.2	Construal of Grouping Principles in Text	26
2.3	Metafunctional Solidarity across Planes	30
2.4	Nominal Construal of One Aspect of Darwin's Theory of Evolution	31
2.5	The Tension between Literal and Figurative Readings of Gould's Nominalization of Darwin's Theory	31
2.6	Stratification: Levels of Abstraction in Language	32
2.7	Language in Relation to its Connotative Semiotics: Ideology, Genre and Register	38
2.8	Constituency Representation for Part–Whole Relations	40
2.9	A Partial System Network for 'Being' Clauses	42
2.10	Experiential and Interpersonal Analysis	43
2.11	The Attributive Clause	43
2.12	Relational-Clauses Paradigm	43
4.1	Interlocking Definitions of Five Technical Terms	72
4.2	Kinds of Climate (Superordination)	73
4.3	Parts of Climate (Composition)	74
5.1	Transitivity (Ideational), Mood (Interpersonal), and Theme and Information (Textual) Structures in the 'Favourite' Clause Type	88
5.2	Motifs Constructed as Theme	97
5.3	Motifs Constructed as New	100
8.1	Vernacular Composition Taxonomy of Body Parts for Bodybuilders	139
8.2	Birdwatchers' Vernacular Taxonomy of Birds of Prey (Raptors)	140
8.3	Uninformed Vernacular Taxonomy of Birds of Prey	141
8.4a	Scientific Taxonomy of Birds of Prey (Falconiformes)	141
8.4b	Scientific Taxonomy of Birds of Prey: Acciptridae	142
8.5	Scientific and Vernacular Taxonomies	143
8.6	Geographical Taxonomy of Climate	144
8.7	Composition Taxonomy of the Ecosystem	156
8.8	Superordination Taxonomy of the Ecosystem	156
8.9	A Generalized Model	159
8.10	How Water Gets into the Air	160

8.11	How Clouds Form	160
8.12	How Rain Falls	160
8.13	How Cloud Droplets Combine — A	161
8.14	How Cloud Droplets Combine — B	161
8.15	Adiabatic Cooling	161
9.1	Organization of Solar Systems: A Scientist's View	169
9.2	Organization of the Universe: A Common-Sense View	170
9.3	Composition Diagram	174
9.4	Classification Diagram	174
9.5	The Structure of an Insect	175
9.6	Classification of Arthropods	176
9.7	Types of Insects	177
9.8	Processes of Change in Geology	178
9.9	A Geologist's Classification of Sedimentary Rocks	180
9.10	A Sea Breeze	181
9.11	Osmosis (Heffernan)	182
9.12	Osmosis (Messel)	184
9.13	The Ear of Man	200
10.1	Taxonomy of English	204
10.2	Common-Sense Taxonomy of Diseases	205
10.3	Medical Taxonomy of Diseases	206
10.4	Taxonomy of Ecosystems: Deserts	207
10.5	Taxonomy of Conservers: Cacti	208
10.6	Taxonomy of Desert Landforms	209
11.1	Taxonomy of Army Women (Text 3)	229
11.2	Taxonomy of Conducting Substances (Text 1)	230
11.3	Relational Processes in English	231
11.4	Conjunctive Relations in Text 2 (Science)	234
11.5	Conjunctive Relations in Text 4 (History)	236
11.6	Text as Wave — Local and Global Patterns of Theme and New	252

Introduction

In *Language as Social Semiotic* (1978), Michael Halliday maps the possibilities of a 'social-functional' approach to linguistics. There he distinguishes 'intra-organistic' from 'inter-organistic' explanations of language, noting the contemporary tendency to explain language use by reference to the 'intra-organistic', that is, by reference to internal, psychological states. This critique applies most obviously to psycholinguistic and cognitive developmental models of 'language acquisition', which have been central in the shaping of literacy in Western schools. But, Halliday indicates, it also applies to those socio-linguistic and ethnographic models that pay homage to 'social context', but tend to explain away the constitutive institutional and ideological relations of text and discourse in terms of individuals' knowledges, intents, states of mind, competences, and so forth. In its stead, Halliday has developed a theory of language as social semiotic performance. The traditions that systemic linguistics draws upon, from Hjelmslev and Malinowski to Durkheim and Voloshinov, are used to build a recognition of the primacy of the social.

Writing Science shows how modern science is, first and foremost, a discourse technology. Here Halliday and James R. Martin show scientific discourse at work in a range of historical, contemporary, and cross-cultural sites: from the works of nineteenth century science, to other cultures' textual representations of the natural world, from school students' writings on scientific knowledge and procedures, to the construction of a 'Secret English' of science in secondary school textbooks and classroom talk. Throughout these essays, science is not taken as a canon of 'great' ideas and truths, nor as a corpus of universal procedures or methods, or, even more mystically, as the product of 'genius', specific mental dispositions and attitudes. Rather, science is conceived of as inter-organistic practice, a linguistic/semiotic practice which has evolved functionally to do specialized kinds of theoretical and practical work in social institutions. Accordingly, Halliday and Martin argue that scientific texts needn't be 'alienating' and 'anti-democratic', but can be deconstructed and made accessible, as part of a broad agenda to linguistically 'construe a world which is recognizable to all those who live in it'. Running across these essays is a commitment not just to remaking science as a humane endeavour, but also to developing new analytic perspectives for critiquing science.

As Derrida (1974) and Lyotard (1982) would insist, to speak of 'science' and a 'science of writing' is to presuppose and build a possible world where both writing and science have been assigned special significance. It should not be surprising, then, that current debates over postmodernity, technology, and attendant shifts in political economy and culture turn on the place of the texts and institutions of science. What counts as 'science' in the period since World War II has been focal in the development of Western nation states, to the point where historical 'winners' and 'losers' in economic, strategic and geopolitical realms are assessed in terms of technological and scientific prowess. The yoking together of scientific work with the imperatives of capital and government had, of course, ample precedents in the late nineteenth century and early twentieth century UK, USA and Germany. But post-war technocratic society had its basis in the symbiotic relationships forged between governments, corporations and research science during World War II (De Landa, 1991). Then, applied approaches to physics, mathematics and statistics, electronics and computing, communications and systems theories, engineering and chemistry emerged — advances which would reshape the character of everyday life and mass consumer culture, research and academic institutions, and indeed, the international distribution of wealth and power. Long before McLuhan and colleagues talked of a transnational culture of electronic signs and semiotic exchange — the necessary material conditions were set in the linking of academic scientific research, first with the state-funded military industrial complexes, and later with the interests of corporate capital. The heady effects of this mix remain, as the Gulf War, economic globalisation and current ethical and legal debates over AIDS research, the Human Genome Project and reproductive technologies, global warming and environmental desecration remind us. What has come to count as science in technocratic culture is the applied, the corporate and the profitable.

In this period of flux and transition, disciplinary and institutional boundaries between science and humanities, between the 'hard' natural sciences and 'soft' human sciences, between the public discourses of science and domains of folk wisdom have become the focus of unprecedented scrutiny. This could be attributed to 'paradigm shift', as attested to by the scepticism within scientific communities towards classical mythologies of scientific objectivity, method and discovery, and the increasing attention paid to the significance of accident, intuition, ignorance and, indeed, chaos in inquiry. But driving the debate has been the sustained sociological and philosophical critique of scientific work, knowledge and power. Social sciences have made the cultural and sociological workings of science an object of study: in the poststructuralist critique of sciences by Foucault and Lyotard; the critique of the 'science of writing' by Derrida; feminist analyses of patriarchy and science by Harding and Haraway; and critical sociologies and ethnohistories of scientific inquiry by Hacking, Rose, Gould, Woolgar and others (for an introduction and bibliography, see Darnovsky, Epstein & Wilson, 1991).

What is at stake is far more than disciplinary and academic turf. In the midst of tenacious political and economic legitimation crises, the whole game may be up for grabs. In a global economy where reliance on technological 'growth' and 'progress' is greater than ever — the power of scientific discourse (and its kin, pseudo-science and pop science) is arguably greater than ever before. The very dependency on corporate science and technological expansion as means for the expansion of state power and legitimacy have translated the crises of economies and cultures into the crises of sciences.

The 'time lag' between such debates and a remaking of science education is not surprising, given the persistent problems of reform of school systems and curriculum largely built by industrial-era design and edict. In the USA and Canada, the UK and Australia, the post-war human capital model of education has proven resilient and recyclable, despite little evidence that it works (Lingard, Porter & Knight, 1992). Faced with service and information-driven economies, finite resources, and shifting patterns and sectors of employment — educational policies and practices in these countries have been driven by the economic promises affiliated with the increased educational 'output' of scientific expertise. Yet this continued mystique about the achievement and value of scientific knowledge stands in contrast to the relative underdevelopment of approaches to science education. Educational research and teacher education have made halting moves from fact-transmission pedagogies to progressivist models which stress exploration and construction of scientific knowledge. Yet apart from Jay Lemke's *Talking Science* (1990), a theoretical and educational companion to *Writing Science*, the reworking of scientific education in terms of language, text and discourse has not received widespread attention from science educators, curriculum developers or teacher educators.

A socially-based linguistic analysis of the texts and discourses of scientific work, then, is an important political and pedagogical move. For many readers, these essays will be a first introduction to systemic functional linguistics at work. In Halliday's model, lexicogrammatical choice can be traced systematically to social/ideological function. This puts it at odds with mainstream linguistic analyses, which do not effectively unknot the reflexivity between the social and the semiotic, between context of situation and lexicogrammar.

Beginning from an introduction to key concepts of systemic linguistics, Halliday and Martin here move on to explore the historical relationships between science, language and literacy. There is vigorous academic dispute over the connections between scientific institutions and technologies of inscription, but historians and anthropologists from Mumford to Goody insist that writing is the enabling technology for 'doing' modern science. As many have pointed out, the writing of history and the framing of disciplinary knowledges were closely allied with the formalization of the Greek alphabetic system. Halliday and Martin argue that the languages and discourses of science indeed have characteristic features that have evolved to do various forms of cognitive and semiotic work which the 'common-sense' language of everyday life cannot: including, for instance, the representation of technicality and abstraction. Hence, the answer of many educational and public approaches to science — to do away with 'jargon' and return to 'plain English' — is a naive educational solution, doing justice neither to scientific work and knowledge, nor to those students who require direct access to the registers of disciplinary knowledge in order to progress through academic systems.

Writing Science tables a range of questions which have long been at the periphery of educational and curriculum debate. Partly because higher achievement in the traditional natural sciences has been an exclusive domain of upper and middle class male elite — little systematic effort has been devoted to exploring scientific literacy and an inclusive pedagogy that better enfranchises women, aboriginal peoples, ethnic minorities and working class children. Part II here moves towards a pedagogy which centres on how scientific language and texts work. This approach will be of value for those teaching science at all levels, for those introducing first and second language learners to the 'special purposes' of scientific writing. The linguistic analysis of what Halliday, Martin and colleagues call the 'Secret English' of school science is the basis for an approach to critical literacy which differs significantly from the emphasis on personal voice, identity and expression prevalent in UK and US. The emphasis here is on self-conscious control over text types and their special linguistic features. Halliday and Martin thus see critical science and critical literacy as commensurate and worthy educational aims, part of a move towards 'more democratic forms of discourse'.

There might indeed be scientific 'facts', constructed and contestable but verifiable truth claims about the natural and social worlds. But, as Lyotard (1984) has pointed out, what is crucial is how these are strung together into a narrative syntax, how they are chained together into sequences of agents and causes, relations and consequences. *Writing Science* indicates how the work of science is necessarily grammatical: naming, constructing and positioning the social and natural worlds, and doing so in a way which builds social relationships of power and knowledge between writers and readers. If we view science as discourse performance, not as a mental accomplishment but as social and textual practice — then what is entailed is nothing less than a political economy of discourse. How science is written and read, spoken and heard is necessarily tied up with questions of access, education and critical literacy.

Allan Luke
Townsville, Australia
October, 1992

References

DARNOVSKY, M., EPSTEIN, S. and WILSON, A. (Eds) (1991) 'Radical Experiments: Social Movements Take on Technoscience'. Theme issue of *Socialist Review*, 21 (2).

DE LANDA, M. (1991) *War in the Age of Intelligent Machines*, New York, Zone Books.

DERRIDA, J. (1974/1982) *Of Grammatology*, translated by G.C. SPIVAK, Baltimore, Johns Hopkins University Press.

HALLIDAY, M.A.K. (1978) *Language as Social Semiotic: The Social Interpretation of Language and Meaning*, London, Edward Arnold.

LEMKE, J.L. (1990) *Talking Science: Language, Learning and Values*, Norwood, New Jersey, Ablex.

LINGARD, R., PORTER, P. and KNIGHT, J. (Eds) (1992) *Schooling Reform in Hard Times*, London, Falmer Press.

LYOTARD, J.F. (1984) *The Postmodern Condition: A Report on Knowledge*, translated by B. MASSUMI, Manchester, University of Manchester Press.

Introduction:

The Discursive Technology of Science

Chapter 1

General Orientation

Adults may choose to deny it, but children in school know very well that there is a 'language of science'. They may not be able to say how they know it; but when they are faced with a wording such as:

> One model said that when a substance dissolves, the attraction between its particles becomes weaker. (Junior Secondary Science Project, 1968, pp. 32–33)

they have no trouble in recognizing it as the language of a chemistry book. And they tend to feel rather put off by it, especially when they find themselves challenged with a question like this one. 'What might happen to the forces of attraction which hold the particles of potassium nitrate together' (ibid.)

If children do get put off by this, we respond, as seems natural to us, by giving their feeling a name. We call it 'alienation'. We have now labelled the condition; we think that in labelling it we have diagnosed it, and that in diagnosing it we are half way towards curing it. In reality, of course, we have only made the condition worse. Nothing could be more alienating than to learn that you are suffering from alienation. But in responding in this way we have helped to demonstrate how scientific discourse works.

It is not only schoolchildren who have felt alienated by the discourse of science. Within a century of the so-called 'scientific revolution' in Europe, people were feeling disturbed by the picture that science presented, of a universe regulated by automatic physical laws and of a vast gulf between humanity and the rest of nature. Prigogine and Stengers, in their remarkable book *Order out of Chaos*, show how this feeling arose; and they point to the disturbing paradox between the humanist origins of natural science and its contemporary image as something unnatural and dehumanizing:

> Science initiated a successful dialogue with nature. On the other hand, the first outcome of this dialogue was the discovery of a silent world. This is the paradox of classical science. It revealed to men a dead, passive nature, a nature that behaves as an automaton which, once programmed, continues to follow the rules inscribed in the program. In this sense the dialogue with nature isolated man from nature instead of bringing

him closer to it. A triumph of human reason turned into a sad truth. It seemed that science debased everything it touched. (Prigogine and Stengers, 1984, p. 6)

To understand this paradox, we have to take account of the kind of language in which science is construed. In his revealing account of science education, based on a study he carried out in New York secondary schools, Jay Lemke put it this way:

How does science teaching alienate so many students from science? How does it happen that so many students come away from their contact with science in school feeling that science is not for them, that it is too impersonal and inhuman for their tastes, or that they simply 'don't have a head for science'? One way this happens, I believe, is through the way we talk science. The language of classroom science sets up a pervasive and false opposition between a world of objective, authoritative, impersonal, humourless scientific fact and the ordinary, personal world of human uncertainties, judgments, values, and interests. (Lemke, 1990a, pp. 129–30)

But the language of classroom science is simply the language of science adapted to the classroom. It fails to overcome the problem; but it did not create it in the first place. The issue is that of the discourse of science itself.

Where children are most likely to be put off is in the early years of secondary school, when they first come face to face with the language of their 'subjects' — the disciplines. Here they meet with unfamiliar forms of discourse; and since these often contain numbers of technical terms, when we first reflect on scientific language we usually think of these as the main, perhaps the only, source of the difficulty. There are a lot of technical terms, of course, and they may be quite hard to master if they are not presented systematically. But children are not, on the whole, bothered by technical terms — they often rather enjoy them; and in any case textbook writers are aware of this difficulty and usually manage to avoid introducing too many of them at once. It is not difficult, however, to find passages of wording without many technical terms which are still very clear instances of scientific writing; for example

One property at least (the colour) of the substance produced is different from the substances that were mixed, and so it is fairly certain that it is a **new substance**. (Junior Secondary Science Project, 1968, p. 43)

Compare the example quoted in chapter 2 below:

Your completed table should help you to see what happens to the risk of getting lung cancer as smoking increases. (Intermediate Science Curriculum Study, 1976, p. 59)

And this is not simply a feature of the language of science textbooks; the following extract from the *Scientific American* contains hardly any technical terms:

Our work on crack growth in other solids leads us to believe that the general conclusions developed for silica can explain the strength behaviour of a wide range of brittle materials. The actual crack tip reactions appear to vary from material to material and the chemistry of each solid must be considered on a case-by-case basis. (Michalske and Bunker, 1987, p. 81)

Of course, technical terms are an essential part of scientific language; it would be impossible to create a discourse of organized knowledge without them. But they are not the whole story. The distinctive quality of scientific language lies in the lexicogrammar (the 'wording') as a whole, and any response it engenders in the reader is a response to the total patterns of the discourse.

Naturally it would engender no response at all unless it was a variety of the parent language. Scientific English may be distinctive, but it is still a kind of English; likewise scientific Chinese is a kind of Chinese. If you feel alienated by scientific English this is because you are reacting to it as a form of a language you already know very well, perhaps as your mother tongue. (If on the other hand you are confronting scientific English directly as a second language, you may find it extraordinarily difficult, especially if it is your first encounter with a language of science; but that is very different from being alienated by it.) It is English with special probabilities attached: a form of English in which certain words, and more significantly certain grammatical constructions, stand out as more highly favoured, while others correspondingly recede and become less highly favoured, than in other varieties of the language. This is not to imply that there is one uniform version of it, any more than when we talk of British English or Australian English we are implying that there is one uniform version of each of these dialects. Any variety of a language, whether functional or dialectal, occupies an extended space, a region whose boundaries are fuzzy and within which there can be considerable internal variation. But it can be defined, and recognized, by certain syndromes, patterns of co-occurrence among features at one or another linguistic level — typically, features of the expression in the case of a dialect, features of the content in the case of a functional variety or 'register'. Such syndromes are what make it plausible to talk of 'the language of science'.

Given the view of language that prevails in western thought, it is natural to think of the language of science as a tool, an instrument for expressing our ideas about the nature of physical and biological processes. But this is a rather impoverished view of language, which distorts the relationship between language and other phenomena. The early humanists, founders of modern science in the West, paid more serious attention to language in their endeavours. In part, this was forced upon them because they were no longer using the language that had served their predecessors, Latin, and instead faced the job of developing their various emerging 'national' languages into resources for construing knowledge. But their concern with language went deeper than that. On the one hand they were reacting against what they saw as (in our jargon of today) a logocentric tendency in medieval thought; the best-known articulation of this attitude is Bacon's 'idols of the marketplace' (*idola fori*), one of the four *idola* or false conceptions which he felt distorted scientific thinking. The *idola fori* result, in Dijksterhuis' words, from the thoughtless use of language, from the delusion that there must correspond to all names actually existing things, and from the confusion of the literal and the figurative meaning of a word. (Dijksterhuis, 1961, p. 398)

The 'delusion' referred to here had already been flagged by William of Occam, whose often quoted stricture on unnecessary entities was in fact a warning against reifying theoretical concepts such as 'motion'; the perception that lay behind this suspicion of language was later codified in the nominalist philosophy of John Locke, summed up by David Oldroyd as follows:

The important point, of course, is that the new philosophy claimed that new knowledge was to be obtained by experimentation, not by analysis of language or by establishing the correct definitions of things. If you wanted to know more about the properties of gold than anyone had ever known before you would need a chemical laboratory, not a dictionary! (Oldroyd, 1988, pp. 91–92)

On the other hand, the scholars of the new learning were at the same time extremely aware of how crucial to their enterprise was the role that language had to play. Since 'language' now meant 'languages', the perception of this role differed somewhat from one country to another; it was stated most explicitly in England and in France, partly perhaps because of the historical accident that these languages, which had changed catastrophically in the medieval period, were having more trouble sorting out their orthographies than Italian, German or Dutch. Whatever the reason, English and French scholars devoted much effort to designing a language of science; the work in England is described and evaluated by Vivian Salmon in her book *The Study of Language in Seventeenth Century England*, published in 1979. This work went through several phases, as those concerned progressively refined their conception of what it was that was needed to make their language effective as a resource for the new knowledge.

The earliest effort was simply to devise a form of shorthand, a writing system that would be simpler and more expeditious in codifying knowledge in writing; for example Timothy Bright's *Characterie*, published in 1588. Bright's work however already embodied a second, more substantial aim: that of providing a universal charactery, a system of writing that would be neutral among the various different languages, in the way that numerical symbols are. Bright appreciated the lexigraphic nature of Chinese writing (that its symbols stood for words, or their parts), which had then recently become accessible in Europe, and used that as a model for his purpose.

Within the next few decades, a more ambitious goal was being pursued, that of a universal 'philosophical language': that is, a fully designed, artificial language that would serve the needs of scientific research. Among those who conceived of plans for such a philosophical language, Vivian Salmon refers to William Petty, Seth Ward, Francis Lodowick, George Dalgarno and John Wilkins; it was the last of these who actually carried out such a plan to the fullest extent, in his famous *Essay Towards a Real Character and a Philosophical Language*, published in 1668. Wilkins' impressive work was the high point in a research effort in which scholars from many countries had been deeply engaged, as they worked towards a new conception of the structure and organization of scientific knowledge.

A 'philosophical language' was not simply a means of writing down, and hence transmitting, knowledge that had already been gained; more than that, it was a means of arriving at new knowledge, a resource for enquiring and for thinking with. The ultimate goal in the conception of scientific language design was subsequently articulated by Leibniz, who (in Oldroyd's words) 'envisaged the construction of a general science of symbols which could be applied to experience' — a project, however, which 'remained unfulfilled in Leibniz' time and remains so to this day' (op. cit., pp. 104–5). But from the efforts and achievements of Wilkins and his contemporaries, and in particular from the extent to which the scientists themselves supported and participated in these efforts, we can gain a sense of the significance accorded to language in seventeenth-century scientific thought. Language was an essential component in enlarging the intellectual domain.

The biggest single demand that was explicitly made on a language of science was that it should be effective in constructing technical taxonomies. All natural languages embody their own folk taxonomies, of plants and animals, diseases, kinship structures and the like; but these are construed in characteristically messy ways, because of the need to compromise among conflicting criteria, and they were seen rather as an obstacle to developing the systematic technical taxonomies that were required by the new science. So when the scientists came to design their own artificial languages much of the emphasis was placed on building up regular morphological patterns for representing a classificatory system in words.

Clearly this had to be one of the central concerns. Unlike commonsense knowledge, which can tolerate — indeed, depends on — compromises, contradictions and indeterminacies of all kinds, scientific knowledge as it was then coming into being needed to be organized around systems of technical concepts arranged in strict hierarchies of kinds and parts. In the event, none of the artificial languages was ever used for the purpose; but the experience of linguistic design that went into creating them was drawn upon in subsequent work, for example in constructing systematic nomenclatures for use in botany and in chemistry. Even where no special linguistic structures have been developed for the purpose, an essential feature of all scientific registers since that time has been their systems of technical terms.

But there is another aspect of scientific language that is just as important as its technical terminology, and that is its technical grammar. Interestingly, the seventeenth century language planners paid no attention to this. Wilkins' philosophical language did, of course, incorporate a grammar — otherwise it would not have been a language, in any practical sense; but it was a grammar of a conventional kind, without any of the innovatory thinking that had gone into the lexical morphology. Yet if we examine how scientists such as Newton were actually using language in their own writings, we find innovations in the grammar which are no less striking than those embodied in the construction of technical terms. People are, of course, less conscious of grammar than they are of vocabulary; no doubt this is one reason for the discrepancy. The other reason would have been, perhaps, that the grammatical developments were more gradual; they were just one further move in a steady progression that had been taking place since the time of Thales and Pythagoras in ancient Greece, and they did not involve creating new grammatical forms so much as systematically deploying and extending resources that were potentially already there.

It is convenient to think of the new resources that came into scientific English (and other languages: for example the Italian of Galileo) at this time as falling under these two headings, the lexical and the grammatical. The 'lexical' resources were highly visible, in the form of vast numbers of new technical terms; what was significant, however, was not so much the terms themselves as the potential that lay behind them. On the one hand, as we have seen, they could be formed into systematic taxonomic hierarchies; on the other hand, they could be added to *ad infinitum* — today a bilingual dictionary of a single branch of a scientific discipline may easily contain 50,000–100,000 entries. The 'grammatical' resources were the constructions of nominal groups and clauses, deployed so that they could be combined to construe a particular form of reasoned argument: a rhetorical structure which soon developed as the prototypical discourse pattern for experimental science. Any passage of Newton's writings could be taken to illustrate these resources, both the lexical and the grammatical, for example the following passage taken from the *Opticks*:

> If the Humours of the Eye by old Age decay, so as by shrinking to make the *Cornea* and Coat of the *Crystalline Humour* grow flatter than before, the Light will not be refracted enough, and for want of a sufficient Refraction will not converge to the bottom, of the Eye but to some place beyond it, and by consequence paint in the bottom of the Eye a confused Picture, and according to the Indistinctness of this Picture the Object will appear confused. This is the reason of the decay of sight in old Men, and shews why their Sight is mended by Spectacles. For those Convex glasses supply the defect of Plumpness in the Eye, and by increasing the Refraction make the Rays converge sooner, so as to convene distinctly at the bottom of the Eye if the glass have a due degree of convexity. And the contrary happens in short-sighted Men whose Eyes are too plump. (Newton, 1704, pp. 15–16)

This is not the place to discuss such language in detail; the relevant features are taken up specifically in the chapters that follow. But we can illustrate the two sets of resources referred to above. Lexically, expressions such as **Crystalline Humour** (here shown to be a kind of **Humour**), **Refraction** (defined earlier in association with **Reflexion**), **Convex** and **convexity** (contrasted a few lines further down with the **Refractions diminished by a Concave-glass of a due degree of Concavity**) are clearly functioning as technical terms. Grammatically, a pattern emerges in which an expression of one kind is followed shortly afterwards by a related expression with a different structural profile:

> will not be refracted enough . . . for want of a sufficient Refraction
> paint(.) a confused Picture . . . according to the Indistinctness of this Picture
> make the Cornea (.) grow flatter . . . supply the defect of Plumpness in the Eye
> those Convex glasses . . . if the Glass have a due degree of convexity.

In each of these pairs, some verb or adjective in the first expression has been reworded in the second as a noun: **refracted — Refraction, confused —**

Indistinctness, [grow] flatter — [the defect of] Plumpness, Convex — convexity; and this has brought with it some other accompanying change, such as **will not be refracted enough — for want of a sufficient Refraction, a confused Picture — the Indistinctness of this Picture**. In each case a grammatical process has taken place which enables a piece of discourse that was previously presented as new information to be re-used as a 'given' in the course of the succeeding argument.

But when we observe these two features, technical vocabulary and nominalized grammar, in a passage of scientific text — even a very short extract like the one just cited — we can see that they are interdependent. Creating a technical term is itself a grammatical process; and when the argument is constructed by the grammar in this way, the words that are turned into nouns tend thereby to become technicalized. In other words, although we recognize two different phenomena taking place (as we must, in order to be able to understand them), in practice they are different aspects of a single semiotic process: that of evolving a technical form of discourse, at a particular 'moment' in socio-historical time.

There is no mystery about this being, at one and the same time, both one phenomenon and two. When we look at it from the standpoint of the wording — lexicogrammatically, or 'from below' in terms of the usual linguistic metaphor — it involves two different aspects of the language's resources, one in the word morphology, the other in the syntax. When we look at it from the standpoint of the meaning — semantically, or 'from above' — we see it as a single complex semogenic process. Lexicogrammatically, it appears as a syndrome of features of the clause; semantically, it appears as a feature of the total discourse. To get a rounded picture, we have to be able to see it both ways.

Here we can, obviously, offer no more than a while-you-wait sketch of one facet of the language of science — although an important one; but it will be enough, perhaps, to enable us to take the next step in our own argument. The language of science is, by its nature, a language in which theories are constructed; its special features are exactly those which make theoretical discourse possible. But this clearly means that the language is not passively reflecting some pre-existing conceptual structure; on the contrary, it is actively engaged in bringing such structures into being. They are, in fact, structures of language; as Lemke has expressed it, 'a scientific theory is a system of related meanings'. We have to abandon the naïve 'correspondence' notion of language, and adopt a more constructivist approach to it. The language of science demonstrates rather convincingly how language does not simply correspond to, reflect or describe human experience; rather, it interprets or, as we prefer to say, 'construes' it. A scientific theory is a linguistic construal of experience.

But in that respect scientific language is merely foregrounding the constructive potential of language as a whole. The grammar of every natural language — its ordinary everyday vocabulary and grammatical structure — is already a theory of human experience. (It is also other things as well.) It transforms our experience into meaning. Whatever language we use, we construe with it both that which we experience as taking place 'out there', and that which we experience as taking place inside ourselves — and (the most problematic part) we construe them in a way which makes it possible to reconcile the two.

Since we all live on the same planet, and since we all have the same brain capacities, all our languages share a great deal in common in the way experience is construed. But within these limits there is also considerable variation from one language to another. Prigogine and Stengers remark, in the Preface to the English translation of their book:

> We believe that to some extent every language provides a different way of describing the common reality in which we are embedded. (Prigogine and Stengers, 1984, p. 31)

— and they are right. Much of this variation, however, is on a small scale and apparently random: thus, the minor differences that exist between English and French (the language in which their book was originally written), while irritating to a learner and challenging to a translator, do not amount to significantly different constructions of the human condition. Even between languages as geographically remote as English and Chinese it is hard to find truly convincing differences — perhaps the gradual shift in the construction of time from a predominantly linear, past/future model at the western end of the Eurasian continent (constructed in the grammar as tense) to a predominantly phasal, ongoing/terminate model at its eastern end (constructed in the grammar as aspect) would be one example, but even there the picture is far from clear. By and large there is a fair degree of homogeneity, in the way our grammars construe experience, all the way from Indonesia to Iceland.

This is not really surprising. After all, human language evolved along with the evolution of the human species; and not only along with it but as an essential component in the evolutionary process. The condition of being human is defined, *inter alia*, by language. But there have been certain major changes in the human condition, changes which seem to have taken place because, in some environments at least, our populations tend inexorably to expand (see for example Johnson and Earle, *The Evolution of Human Societies*; on p. 16 they sum up their findings by saying 'The primary motor for cultural evolution is population growth'). The shift from mobility (hunting and gathering) to settlement (husbanding and cultivating) as the primary mode of subsisting was one such catastrophic change; this may have been associated with quite significant changes in the way experience is construed in language. The classic statement on this issue was made by Benjamin Lee Whorf, in his various papers collected under the title *Language, Thought and Reality*; despite having been 'refuted' many times over this remains as viable as it was when it was first written. More recently, Whorf's ideas have come to be discussed with greater understanding, e.g., by Lee, 1985, Lucy, 1985, and Lucy and Wertsch, 1987.

It would be surprising if there were not some pervasive differences in world view between two such different patterns of human culture. Since some sections of humanity have continued to pursue a non-settled way of life, it ought to be possible to compare the language of the two groups; but this has still not been satisfactorily achieved — for two main reasons. One is that the random, local variation referred to earlier gets in the way; if we focus on grammatical structure, then all types of language will be found everywhere, but it is the underlying 'cryptotypic' grammar that would vary in systematic ways, and we have hardly begun to analyse this. The other reason is that many linguists have felt discouraged by the risk of being attacked as naïve historicists (at best, and at worst as racists if they ventured to suggest any such thing. But to recognize that the changeover from mobility to settlement, where it took place, was an irreversible process is not in any way to attach value to either of these forms of existence. The point is important in our present context, because there appear to have been one, or perhaps two, comparably significant changes in the course of human history, likewise involving some populations and not others; and the 'scientific revolution' was one. (The other, perhaps equally critical, was the 'technological revolution' of the Iron Age.)

Let us be clear what we are saying here. It is not in dispute that, for whatever reason, certain human societies evolved along particular lines following a route from mobility to settlement; among those that settled, some evolved from agrarian to technological, and some of these again to scientific-industrial. The question we are asking is: What part does language play in these fundamental changes in the relationship of human beings to their environment? One answer might be: none at all. It simply tags along behind, coining new words when new things appear on the scene but otherwise remaining unaffected in its content plane (its semantics and its grammar). In this view, any changes that took place in language were merely random and reversible, like the changes from one to another of the morphological language 'types' set up in the nineteenth century (isolating, agglutinative, inflexional).

We reject this view. In our opinion the history of language is not separate from the rest of human history; on the contrary, it is an essential aspect of it. Human history is as much a history of semiotic activity as it is of socio-economic activity. Experience is ongoingly reconstrued as societies evolve; such reconstrual is not only a necessary condition for their evolution — it is also an integral part of it. We have barely started to understand the way this happens (cf. Lemke, 1992); partly because, as already stressed, our descriptions of languages are not yet penetrating enough, but also because we do not yet fully comprehend how semiotic systems work. (We shall come back to this point below.) But while we may not yet understand how meaning evolved, this is no reason for denying that it did evolve, or for assuming that all semantic systems were spontaneously created in their present form.

When we come to consider a special variety of a language, such as the language of science, we may be better able to give some account of how this evolved; not only has it a much shorter history, but also we can assume that whatever special features it has that mark it off from other varieties of the language have some particular significance in relation to their environment. Or rather, we can assume that they had some particular significance at the time they first appeared; it is a common experience for such features to become ritualized over the course of time, once the social context has changed, but it is virtually certain that they would have been functional in origin. We shall try to show, in the chapters which follow, something of the extraordinarily complex ideological edifices which are constructed and maintained by scientific discourse, and how the grammar has evolved to make this discourse possible. We shall also try to show how, in the process, the grammar has been reconstruing the nature of experience.

It is not too fanciful to say that the language of science has reshaped our whole world view. But it has done so in ways which (as is typical of many historical processes) begin by freeing and enabling but end up by constraining and distorting. This might not matter so much if the language of science had remained the special prerogative of a priestly caste (such a thing can happen, when a form of a language becomes wholly ceremonial, and hence gets marginalized). In our recent history, however, what has been happening is just the opposite of this. A form of language that began as the semiotic underpinning for what was, in the worldwide context, a rather esoteric structure of knowledge has gradually been taking over as the dominant mode for interpreting human existence. Every text, from the discourses of technocracy and bureaucracy to the television magazine and the blurb on the back of the cereal packet, is in some way affected by the modes of meaning that evolved as the scaffolding for scientific knowledge.

In other words, the language of science has become the language of literacy. Having come into being as a particular kind of written language, it has taken over as model and as norm. Whether we are acting out the role of scientist or not, whenever we read and write we are likely to find ourselves conjured into a world picture that was painted, originally, as a backdrop to the scientific stage. This picture represents a particular construction of reality; as Prigogine and Stengers remind us,

> Whatever *reality* may mean, it always corresponds to an active intellectual construction. The descriptions presented by science can no longer be disentangled from our [i.e., the scientists'] questioning activity. (Prigogine and Stengers, p. 55)

But it is a picture that is far removed from, and in some ways directly opposed to, the 'reality' of our ordinary everyday experience. Of course, this too is a construct; it is constructed in the grammar of the ordinary everyday language — the 'mother tongue' that first showed us how the world made sense. But that simply makes it harder for us to accept a new and conflicting version. If you feel that, as a condition of becoming literate, you have to reject the wisdom you have learnt before, you may well decide to disengage. The 'alienation' that we referred to at the beginning is in danger of becoming — some might say has already become — an alienation from the written word.

In the chapters which follow, we have tried to present a rounded picture of the language of science the way it has evolved as a variety of present-day English. In Part 1, the perspective is that of a user: that is, scientific English is treated as something 'in place', and then explored both as system (looked at historically) and as process ('at work' in scientific texts). In Part 2, the perspective is that of a learner: the language is presented as something to be mastered, and then explored as a resource for constructing knowledge and achieving control.

In each of these perspectives, we have given prominence to the lexico-grammatical characteristics of scientific writing. We should therefore make it clear that, in concentrating on the grammar, we are not excluding from the picture the generic aspects of scientific discourse; questions of genre are clearly significant (and are in fact taken up in Part 2). The structure of a scientific paper was explicitly debated by the founders of the Royal Society in London; and although ideas have changed about what this structure should be, editors of journals have always tended to impose their rather strict canons of acceptable written presentation, as regards both the textual format and (more recently also) the interpersonal style. But this aspect of scientific discourse has been rather extensively treated (for example in Charles Bazerman's book *Shaping Written Knowledge*); whereas almost no attention has been paid to the distinctive features of its grammar. Yet it is the grammar that does the work; this is where knowledge is constructed, and where the ideological foundations of what constitutes scientific practice are laid down.

The evolution of science was, we would maintain, the evolution of scientific grammar. We do not mean by this scientific theories of grammar — a scientific 'grammatics'; we mean the grammatical resources of the natural languages by which science came to be construed. In case this seems far-fetched, let us make the point in two steps. The evolution of science was the evolution of scientific thought. But thought — not all thought, perhaps, but certainly all systematic thought — is construed in language; and the powerhouse of a language is its grammar. The process was a long and complex one, and it has hardly yet begun to be seriously researched; but we can try, very briefly, to identify some of the milestones along the way. We shall confine our account to western science, because it was in the west that the move from technology into science first took place; but it should be remembered that the original languages of technology evolved more or less simultaneously in the three great iron-age cultures of China, India and the eastern Mediterranean.

As a first step, the early Greek scientists took up and developed a particular resource in the grammar of Greek, the potential for deriving from the lexical stem of one word another word of a different class (technically, the transcategorizing potential of the derivational morphology). Within this, they exploited the potential for transforming verbs and adjectives into nouns. In this way they generated ordered sets of technical terms, abstract entities which had begun as the names of processes or of properties, like **motion**, **weight**, **sum**, **revolution**, **distance** — or in some cases as the names of relations between processes, like **cause**. Secondly, these scholars — and more specifically the mathematicians — developed the modifying potential of the Greek nominal group; in particular, the resource of extending the nominal group with embedded clauses and prepositional phrases. In this way they generated complex specifications of bodies and of figures; these functioned especially as variables requiring to be measured, for example **the square on the hypotenuse of a right-angled triangle**. As in English (where the structure of the nominal group is very similar) this device was applicable recursively; its semogenic power can be seen in mathematical expressions such as the following from Aristarchus of Samos:

> The straight line subtending the portion intercepted within the earth's shadow of the circumference of the circle in which the extremities of the diameter of the circle dividing the dark and the bright portions in the moon move . . . (Heath, 1913, p. 393)

These resources were then taken over by calquing (systematic translation of the parts) into Latin — without much difficulty, since the two languages were related and reasonably alike (although the second step was slightly problematic because the Latin nominal group was less inclined to accommodate prepositional phrases).

More than anything else, these two potentials of the grammar, that for turning verbs or adjectives into nouns, and that for expanding the scope of the nominal group — including, critically, the potential of combining the two together — opened up a discourse for technology and the foundations of science. In Byzantium, where Greek remained the language of learning, this discourse was eventually absorbed into Arabic, which had itself meanwhile emerged independently as a language of scholarship. In western Europe, where Latin took over, it continued to evolve into medieval times; by then, however, while the outward form was still Latin, the underlying semantic styles were those of the next generation of spoken languages, Italian, Spanish, French, English, German and so on, and further developments, even if first realized in Latin, were more an extension of these languages than of Latin itself. Probably the main extension of the grammar that took place in medieval Latin was in the area of relational processes (types of 'being'), which construed systems of definitions and taxonomies of logical relationships.

Early examples in English of the language of medieval technology and science can be found in Chaucer's *Treatise on the Astrolabe* and *Equatory of Planets*. For scientific English these serve as a useful point of departure. If one then compares the language of these texts with that of Newton, one can sense the change of direction that is being inaugurated in Newton's writing, where the grammar undergoes a kind of lateral shift that leads into 'grammatical metaphor' on a massive scale. Examples such as those given earlier, where e.g., **will not be refracted enough** is picked up by **for want of a sufficient Refraction**, are seldom found in the earlier texts. Expressions such as **a sufficient Refraction** and **the indistinctness of this Picture**, each by itself so slight as to be almost unnoticeable, foreshadow a significant change of orientation in the discourse.

Why do we say these constitute a grammatical metaphor? Because a process, that of 'refracting', which was first construed as a verb (the prototypical realization of a process), then comes to be reconstrued in the form of a noun (the prototypical realization of a thing). The second instance is metaphorical with respect to the first, in the same way that the shift from **imagination** to **painted scenes and pageants of the brain** (from Abraham Cowley, quoted by Peter Medawar, 1984 in *Pluto's Republic*, p. 48) is metaphorical. Here, 'the faculty of producing mental images' is first represented 'literally' as **imagination** and then re-represented as **pageants** which literally represents 'elaborate colourful parades'; this is metaphor in its regular, lexical sense. In grammatical metaphor, instead of a lexical transformation (of one word to another) the transformation is in the grammar — from one class to another, with the word (here the lexical item **refract-**) remaining the same. In the same way, in **a confused Picture . . . the Indistinctness of this Picture** a property, 'unclear', which was first construed as an adjective (the prototypical realization of a property), likewise comes to be reconstrued as a noun. (Here there happens to be also a lexical change, from **confused** to **indistinct**; but this does not involve any further metaphor.)

Now of course there has always been grammatical metaphor in language, just as there has always been lexical metaphor; the original derivations of nouns as technical terms in ancient Greek were already in this sense metaphorical. But there are certain significant differences between these and the later developments. The earlier process was one of transcategorization within the grammar; the meaning construed in this way is a new technical abstraction forming part of a scientific theory, and its original semantic status (as process or property) is replaced by that of an abstract theoretical entity — thus **motion** and **distance** are no longer synonymous with **moving** and **(being) far**. This semogenic process did not of course come to an end; it continues with increasing vigour — it is hard to guess how many new technical terms are created in English each day, but it must amount to quite a considerable number. But in the later development the nominalized form is not in fact being construed as a technical term; rather, it is a temporary construct set up to meet the needs of the discourse, like **plumpness** or **indistinctness**, which still retain their semantic status as properties. We can think of instances like these, of course, as being technicalized for the nonce, and such 'instantial' technicalizations may in time evolve into technical terms; but there is still a difference between the two. This difference can be seen in our first example, where **refraction** is being used not in its role as a technical term of the theory but as a metaphorical nominalization of the verb **refracted** — and so brings with it a little cluster of other grammatical metaphors, whereby the expression of the degree of the process, construed (in prototypical fashion) adverbially as **not . . . enough**, is reconstrued metaphorically as a noun **want** plus an adjective **sufficient** modifying the metaphorized process **refraction** (**for want of a sufficient Refraction**, where **Refraction** is still referring to the process of being refracted).

Such a small beginning may hardly seem worthy to be mentioned. But there is a steady, unbroken evolution in scientific English from this small beginning to the kinds of wording which are typical of written science today:

> A further consequence of the decreasing electronegativity down Group VII is that the relative stability of the positive oxidation states increases with increasing relative atomic mass of the halogen. (Hill and Holman, 1978, p. 243)

> Let us imagine a hypothetical universe in which the same time-symmetric classical equations apply as those of our own universe, but for which behaviour of the familiar kind (e.g., shattering and spilling of water glasses) coexists with occurrences like the time-reverses of these. (Penrose, 1989, p. 397)

> The subsequent development of aerogels, however, was most strongly promoted by their utility in detectors of Cerenkov radiation. (Fricke, 1988, p. 93)

Here the effect of grammatical metaphor can be clearly seen, in expressions such as: the development of x was promoted by their utility in y (less metaphorically, x were developed because they could be used in y); behaviour of the familiar kind, e.g., x-ing of y, coexists with occurrences like z-w (less metaphorically, things behave not only in familiar ways, like y x-ing, but also in ways where the w is z-ed); the relative stability of x increases with increasing y of z (less metaphorically, x becomes more stable as z acquires more y). These examples were not drawn from academic journals; they were taken from randomly opened pages of a senior secondary textbook, a book written for non-specialists, and an issue of the *Scientific American*. Articles written for specialists typically display a considerably denser concentration of grammatical metaphor, which reaches an extreme in the abstracts that areprovided at the beginning.

The birth of science, then (if we may indulge in a well-worn lexical metaphor), from the union of technology with mathematics, is realized semiotically by the birth of grammatical metaphor, from the union of nominalization with recursive modification of the nominal group. This emerging variety of what Whorf called 'Standard Average European', instantiated for example in Galileo's Italian and in Newton's English (in reality, of course, a far more complex construction than this brief sketch can hope to suggest), provided a discourse for doing experimental science. The feature we have picked out as salient was one which enabled complex sequences of text to be 'packaged' so as to form a single element in a subsequent semantic configuration.

But by the same token, something else was also happening at the same time. When wordings are packaged in this way, having started off as (sequences of) clauses, they turn into nominal groups, like **the subsequent development of aerogels** nominalized from **aerogels (were) subsequently developed**. It is this nominalization that enables them to function as an element in another clause. But it also has another effect: it construes these phenomena as if they were things. The prototypical meaning of a noun is an object; when **stable, behave, occur, develop, useful** are regrammaticized as **stability, behaviour, occurrence, development, utility** they take on the semantic flavour of objects, on the model of the abstract objects of a technical taxonomy like **radiation, equation** and **mass**. Isolated instances of this would by themselves have little significance; but when it happens on a massive scale the effect is to reconstrue the nature of experience as a whole. Where the everyday 'mother tongue' of commonsense knowledge construes reality as a balanced tension between things and processes, the elaborated register of scientific knowledge reconstrues it as an edifice of things. It holds reality still, to be kept under observation and experimented with; and in so doing, interprets it not as changing with time (as the grammar of clauses interprets it) but as persisting — or rather, persistence — through time, which is the mode of being of a noun.

This is a very powerful kind of grammar, and it has tended to take over and become a norm. The English that is written by adults, in most present-day genres, is highly nominalized in just this way. Discourse of this sort is probably familiar to all of us:

Key responsibilities will be the investment of all domestic equity portfolios for the division and contribution to the development of investment strategy. (*Sydney Morning Herald*, 1 February 1992, p. 32)

But whereas this nominalizing was functional in the language of science, since it contributed both to technical terminology and to reasoned argument, in other discourses it is largely a ritual feature, engendering only prestige and bureaucratic power. It becomes a language of hierarchy, privileging the expert and limiting access to specialized domains of cultural experience.

Lemke characterizes a language as a dynamic open system: a system that is not stable, but is metastable, able to persist through time only by constantly changing in the course of interacting with its environment. One way in which a language typically changes is that new registers or functional varieties evolve along with changing historical conditions. The evolution of a register of science is a paradigm example of this.

The 'scientific revolution' took place in the context of the physical sciences; it was here that the new conception of knowledge was first worked out. Thus the leading edge of scientific language was the language of the physical sciences, and the semantic styles that evolved were those of physical systems and of the mathematics that is constructed to explain them. This discourse was then extended to encompass other, more complex kinds of system: first biological, then social systems. In calling these 'more complex', we are obviously not comparing them in terms of some overall measure of complexity; we are referring specifically to their relationship to each other as classes of phenomena. A physical system, at least as construed in classical Newtonian physics, is purely physical in nature; but a biological system is both biological and physical, while a social system is at once all three. Hence it was progressively more difficult to understand the kind of abstraction that was involved in construing these various systems: a 'biological fact' is more problematic than a 'physical fact', and a 'social fact' is more problematic still. To put this in other terms, the relationship of an observable instance to the underlying system changes with each step; and the grammar, which developed around the semantics of a physical fact, has to come to terms with, and to naturalize, each of these new types of instantiation.

What the grammar does, as we have seen, is to construe phenomena of any kind into a scientific theory. While there is some minor variation among the different languages in the way this is typically done — for example between English and French, where the former constructs reality more along empiricist, the latter more along rationalist lines — the grammar of scientific theory is largely in common. But what kind of a system is a scientific theory? A theory is a system of yet another kind — a semiotic system. A semiotic system is a system of the fourth order of complexity: that is, it is at the same time physical and biological and social.

The most general case of a semiotic system is a language (in the prototypical sense of this term: a natural language, spoken by adults, and learnt as a mother tongue). This involves a physical medium (typically sound waves), a biological organism as transmitter/receiver, and an interactive social order. Such a system constitutes, as we have expressed it, a general theory of experience: with it we construe our commonsense knowledge of the world, the 'reality' that lies about us and inside us. But construing organized knowledge, in the shape of a scientific theory, means evolving a dedicated semiotic system: a special register of a language which will be orthogonal to, and at the same time a realization of, a system (or rather a universe of systems) of one or more of these four kinds.

It seems to have taken two or three generations for people to come to terms with each new kind of system. If we use the century as a crude but convenient peg, we can say that physical systems were interpreted in the seventeenth and eighteenth centuries, biological systems in the nineteenth century and social systems in our own twentieth century. Of course, scholars had always been thinking about systems of the more complex kinds, and had tried to account for their special characteristics; already, among the ancient Greek scholars, the Stoics had recognized the need for a special theory of the sign to account for semiotic systems such as language. But in the main currents of thought the natural strategy was to map the more complex system on to a kind that is well understood. Thus the modern period language was modelled first as matter, then as matter plus life; until in the early twentieth century Saussure imported from sociology the concept of value. Since a language is a phenomenon of all these kinds, it was possible to learn a great deal about it; but what was learnt did not yet amount to a science of language, because the special nature of semiotic systems had not yet been understood. Language has a fourth sphere of action, one that lies beyond those of matter, of life and of value; it has **meaning**. The unique property of semiotic systems is that they are systems of meaning.

Meaning arises when a dynamic open system of the social kind — one based on value — becomes stratified. Stratification is the feature that was first adumbrated in the classical theory of the sign; the technical name for the relationship that is brought about by stratifying is **realization**. We discuss below how the concept of realization may best be construed in a theory of language, given that we still understand relatively little about it. Lemke has suggested that it may be formalized through the notion of 'metaredundancy' (1984) as the analogue of the cause and effect of a classical physical system. But it is widely misinterpreted; nearly a century after Saussure there are still those who treat it as if it was a relation of cause and effect, asking about a a stratal relationship such as that between semantics and lexicogrammar the naïve question 'which causes which?'. Realization is a relationship of a very different kind, more akin to that of token and value, where the two can only come into existence, and remain in existence, simultaneously.

Linguists often notice how, when highly sophisticated thinkers from other sciences turn their attention to language, they often ignore altogether the findings of linguistics and regress to treating language at the level at which it is presented in the early years of secondary school. We agree that this is a pity; but we are inclined rather to seek the reason why they do it. To us it seems that this happens because they consider that linguistics has not yet evolved into a science; in the formulation we used earlier, the nature of a 'semiotic fact' is still not properly understood. In our view the twentieth-century scholar who came nearest to this understanding was Hjelmslev, with significant contributions from Whorf, from Firth and from Trubetzkoy; one of the few who have tried to build on Hjelmslev's work is Sydney Lamb. Chomsky oriented linguistics towards philosophy, where it had been located for much of its earlier history; but that did not turn it into a science. As one of the leaders of contemporary linguistics, Claude Hagège, has pointed out, it is the working practices of scientists — how they construct theories to explain the phenomena of experience — that provide the model for those (including linguists) who want to 'do science', rather than philosophers' interpretations of these, which are theories constructed to explain how scientists work.

There is no virtue in doing science for its own sake; and in any case linguistics does not become 'scientific' by slavishly following the methods of the physical or other sciences. If the semiotic sciences do develop alongside the others this will change our conception of what 'doing science' means. It will not change the principles of theory construction, or the essentially public nature of scientific activity; but it will add a new type of instantiation, and hence a new relation between the observer and the phenomenon, which will broaden our conception of possible kinds of reality. At present, because the relation between observable instance and underlying system is obscure there is a huge gulf in linguistics between the study of language and the study of text; and this is of practical significance, in that it adversely affects all forms of activity involving language, whether in language education, language pathology or language planning. In this respect, at least, there would seem to be room for a more 'scientific' approach.

Clearly whatever limitations there are to our understanding of language as a whole apply also to our understanding of the language of science. We have tried in the chapters in this book to close the gap between the system and the instance — which are, in fact, different observer positions, not different phenomena — and to interpret the language of science both as system, or potential, and as instantiated in text. We find it helpful, in this context, to locate it in its historical dimension — or rather, its historical dimensions, because a semiotic system moves along three distinct axes of time. First there is phylogenetic time: the system itself evolves, and here 'system' may refer to human language as a whole, to a specific language such as English, or to a specific variety of a language, like scientific English. Secondly, there is ontogenetic time: the language of each human being grows, matures and dies. Thirdly there is what we might call 'logogenetic' time, using *logos* in its original sense of 'discourse': each text unfolds, from a beginning through a middle to an end.

We confront all of these histories when we come to explain grammatical metaphor in the language of science. Given a pair of variants like **(how/that) aerogels subsequently developed** and **the subsequent development of aerogels**, if we view them synoptically all we can say is that each one is metaphorical from the standpoint of the other. But if we view them as related in time, then there is a clear temporal priority. The clausal variant precedes the nominal one in all three dimensions of history: it evolved earlier in the English language (and probably in human language as a whole); it appears earlier in life, as children develop their mother tongue; and it tends to come at an earlier point in the unfolding of a particular text. It is these dynamic considerations that lead us to call the nominal variant metaphorical.

Can we discern any general historical trends that are relevant to the language of science? It seems to us that two things are happening that may influence the way the language of science goes on to evolve in the future. One is that semiotic processes are all the time becoming relatively more prominent in human life in general; the other is that systems of other kinds are coming to be interpreted more and more in semiotic terms. In both these developments language is at the centre, and in particular the language of systematized knowledge. However, it also seems to us that in both these contexts this language is likely to change, and to change in a particular direction. Language is, as we have tried to suggest, both a part of human history and a realization of it, the means whereby the historical process is construed. This is what we mean by language as 'social semiotic': while it accommodates endless random variation of a local kind, in its global evolution it cannot be other than a participant in the social process.

It is a truism to say that we are now in the midst of a period of history when people's lifestyles are changing very fast. With our late twentieth-century technology, many of us no longer spend our time producing and exchanging goods and services; instead, we produce and exchange information. The hub of a city of the industrial revolution was its railway station or its airport, where people and their products were moved around; that of a twenty-first-century city — a 'multifunction polis', as it has been ineptly named — will be (we are told) its information centre, or teleport. The citizens of Osaka, who regard their city as the technological capital of the world (what Osaka thinks today Tokyo will think tomorrow), call it an 'information-oriented international urban complex of the twenty-first century'; its teleport will be

> an information communication base integrating satellite and overland optical fibre network communications systems; it is a port of information communication. (Osaka Port and Harbour Bureau, 1987, p. 7)

In this sort of environment, people will be interfacing more and more with semiotic systems and less and less with social, biological or physical ones — a way of life that is familiar to many human beings already.

As a concomitant of this, scientists are increasingly using semiotic models to complement their physical and biological models of the universe. This began with relativity, as David Bohm makes clear:

> A very significant change of language is involved in the expression of the new order and measure of time plied [*sic*] by relativistic theory. The speed of light is taken not as a possible speed of an *object*, but rather as the maximum speed of propagation of a *signal*. Heretofore, the notion of signal has played no role in the underlying general descriptive order of physics, but now it is playing a key role in this context.
>
> The word 'signal' contains the word 'sign', which means 'to point to something' as well as 'to have significance'. A signal is indeed a kind of *communication*. So in a certain way, significance, meaning, and communication become relevant in the expression of the general descriptive order of physics (as did also information, which is, however, only a *part* of the content or meaning of a communication). The full implications of this have perhaps not yet been realized, i.e., of how certain very subtle notions of order going far beyond those of classical mechanics have tacitly been brought into the general descriptive framework of physics. (Bohm, 1980, p. 123)

Many physical, chemical and biological phenomena are coming to be interpreted as semiotic events. Prigogine and Stengers give the example of periodic chemical processes ('chemical clocks') that occur in far-from-equilibrium states of matter:

> Suppose we have two kinds of molecules, 'red' and 'blue'. Because of the chaotic motion of the molecules, we would expect that at a given moment we would have more red molecules, say, in the left part of a vessel. Then a bit later more blue molecules would appear, and so on. The vessel would appear to us as 'violet', with occasional irregular flashes of red or blue. However, this is *not* what happens with a chemical clock; here the system is all blue, then it abruptly changes its colour to red, then again to blue. Because all these changes occur at *regular* time intervals, we have a coherent process.
>
> Such a degree of order stemming from the activity of billions of molecules seems incredible, and indeed, if chemical clocks had not been observed, no one would believe that such a process is possible. To change colour all at once, molecules must have a way to 'communicate'. The system has to act as a whole. We will return repeatedly to this key word, communicate, which is of obvious importance in so many fields,

from chemistry to neurophysiology. Dissipative structures introduce probably one of the simplest physical mechanisms for communication. (Prigogine and Stengers, 1984, pp. 147–8)

Here 'communicate' is picked out as a 'key word', a word that is 'of obvious importance in so many fields'. But this, in fact, is where we have to demur. The word 'communicate' in itself is of very little importance; nor is the fact that 'the word "signal" contains the word "sign"'', whatever that 'contains' is taken to mean. What is important is the system of meanings that constitute a scientific theory of communication (that is, of semiotic systems and processes), and the lexicogrammatical resources (the 'wordings' as a whole) by which these meanings are construed.

And here we come to a problem and a paradox. The problem is this. The language of science evolved in the construal of a special kind of knowledge — a scientific theory of experience. Such a theory, as we have said, is a semiotic system; it is based on the fundamental semiotic relation of realization, inhering in strata or cycles of token and value. But this means that scientists have all along been treating physical and biological processes as realizations — and hence as inherently communicative (Prigogine and Stengers refer to science as 'man's dialogue with nature'). (We are using 'system' always as a shorthand term for 'system-and-process'; communicating is simply semiotic process.) The problem, now that semiotic systems are being explicitly invoked as explanatory models in science, is to direct the beam of scientific enquiry on to such systems and study them as phenomena in their own right. They can hardly serve an explanatory role if they are not themselves understood.

The prototype of a semiotic system is, as we have said, a natural language; and this leads us in to the paradox. In adapting natural languages to the construction of experimental science, the creators of scientific discourse developed powerful new forms of wording; and these have construed a reality of a particular kind — one that is fixed and determinate, in which objects predominate and processes serve merely to define and classify them. But the direction of physics in the twentieth century has been exactly the opposite: from absolute to relative, from object to process, from determinate to probabilistic, from stability to flow. Many writers have been aware of the contradiction that this has brought about, and have hoped somehow to escape from it by redesigning the forms of language — without realizing, however, that it is not language as such, but the particular register of scientific language, that presents this overdeterminate face. The language they learnt at their mothers' knees is much more in harmony with their deepest theoretical perceptions.

So while there is no reason to doubt that the language of science, as a variety of present-day English (and its counterpart in other languages), will continue to evolve in the twenty-first century, we may expect that it will change somewhat in its orientation. It is likely to shift further towards semiotic explanations, both at the highest level of scientific abstraction and at the technological level in line with the 'information society' (the vast output of computer documentation has already constituted a special sub-register at this level). But at the same time it is likely to back off from its present extremes of nominalization and grammatical metaphor and go back to being more preoccupied with processes and more tolerant of indeterminacy and flux.

In order to do this while still functioning at the technical and abstract level of scientific discourse the grammar would need to be restructured in significant ways. This would not be a matter of inventing a few new verbs; it would mean recasting the nominal mode into a clausal one while developing the verbal group as a technical resource. Note in this connection Whorf's observation about Hopi:

> Most metaphysical words in Hopi are verbs, not nouns as in European languages . . . Hopi, with its preference for verbs, as contrasted with our own liking for nouns, perpetually turns our propositions about things into propositions about events. (Whorf, 1950, pp. 61–63)

It is doubtful whether this could be done by means of design; a language is an evolved system, and when people have tried to design features of language they have almost always failed — although it has to be said that they have usually done so without knowing much about what language is or how it works. But however it came about, any change of this kind would have important social consequences, because it would help to lessen the gap between written language and spoken, and between the commonsense discourse of home and neighbourhood and the elaborated discourse of school and the institutions of adult life.

Two other factors seem to tend in the same direction. One is the way that information technology has developed. The semotechnology of the scientific revolution was print; this made the written language predominant, and greatly exaggerated the difference between writing and speech. Eventually the status of writing was undermined by speech technology — telephone and radio; this redressed the balance somewhat but did not bring the two closer together. The disjunction is being overcome, however, by tape recorders and computers: spoken language can now be preserved through time as text, while written language can be scrolled in temporal sequence up the screen. Instead of artificially forcing the two apart, the new technology tends to mix them up together; as happens for example in electronic mail, which is an interesting blend of spoken and written forms.

But there is another, deeper tendency at work, a long-term trend — however faltering and backtracking — towards more democratic forms of discourse. The language of science, though forward-looking in its origins, has become increasingly anti-democratic: its arcane grammatical metaphor sets apart those who understand it and shields them from those who do not. It is elitist also in another sense, in that its grammar constantly proclaims the uniqueness of the human species. There are signs that people are looking for new ways of meaning — for a grammar which, instead of reconstructing experience so that it becomes accessible only to a few, takes seriously its own beginnings in everyday language and construes a world that is recognizable to all those who live in it. We would then no longer be doomed, as Prigogine and Stengers put it, 'to choosing between an antiscientific philosophy and an alienating science'. (Prigogine and Stengers, 1984, p. 96)

Chapter 2

The Model

Systemic Functional Linguistics

In this book, as already noted, our general approach has been to reason grammatically about the language of science. For this task we have used the theoretical framework of systemic functional linguistics. Systemic functional linguistics (hereafter abbreviated to 'SFL') has been evolved as a tool for participating in political processes. Seeing that the theory and practice of science, and science education, have a central place in our political life, it seems natural for us to adopt a systemic functional perspective.

What are the special features of SFL, as it continues to evolve, that make it effective for an investigation of this kind? It seems to us that it has five orientations that are critical in this respect — and that may be critical in another respect alluded to above, namely that of evolving a scientific discourse for language itself. We will summarize these five orientations here, and then go on to present a brief overview of the systemic functional model.

Rule/resource — To begin, SFL is oriented to the description of language as *a resource for meaning* rather than as a system of rules. It is oriented, in other words, to speakers' meaning potential (what they can mean) rather than neurologically based constraints on what they can say. This orientation has made it easier for us to focus on the semogenesis of scientific discourse, including phylogenesis (evolution in the professional community), ontogenesis (apprenticeship in education) and logogenesis (development in text) — with genesis interpreted as expanding meaning potential.

Sentence/text — Second, SFL is concerned with *texts*, rather than sentences, as the basic unit through which meaning is negotiated. It treats grammar, in other words, as the realization of discourse — from which emerges the conception of a functional grammar, naturally related to its text semantics (as opposed to an autonomous syntax). This concern has made it possible for us to reason grammatically about the semantic organization of scientific texts and the systems of meaning they instantiate.

Text/context — Third, SFL focusses on solidary relations between texts and *social contexts* rather than on texts as decontextualized structural entities in their own right. It looks, in other words, for solidary (i.e., mutually predictive) relationships between texts and the social practices they realize, deliberately sidestepping the question of the role of mental organs in human behaviour — but with semiosis as the resolution of the engagement of physical, biological and social resources (i.e., consciousness) in our species. This focus has encouraged us to shunt between science as institution and science as text, as two complementary perspectives on scientific discourse.

Expressing/construing meaning — Fourth, SFL is concerned with language as a system for construing meaning, rather than as a conduit through which thoughts and feelings are poured. In other words, it views language as a meaning-making system rather than a meaning-expressing one. This concern has made it easier for us to focus on the role of grammar in constructing the uncommonsense interpretation of reality which distinguishes science as a discipline.

Parsimony/extravagance — Finally, SFL is oriented to extravagance, rather than parsimony. It is oriented, in other words, to developing an elaborate model in which language, life, the universe and everything can be viewed in communicative (i.e., semiotic) terms. For us this has meant that there is usually enough descriptive power around for the deconstructive task at hand; and if not, there is room, both theoretical and social, to invent it (cf. Trevarthen, 1992).

It should perhaps be stressed at this point that we do not view our grammatically based deconstructions of text in social contexts (linguistics writ large) as exclusive; quite the contrary, we have found that by broadening our coverage as far as possible in these linguistic terms we have been able to negotiate much more effectively with scholars in related fields than had we sat back and done our bit, working within the confines of a parsimonious model of language as a set of rules — rules defining neurological limitations on the formation of decontextualized sentences (which are arbitrarily related to their meaning and use). The thrust of our work has been transdisciplinary, rather than interdisciplinary, in this respect.

The Model

In the remainder of this introduction we would like to present a brief overview of the systemic functional model of language[1] informing our work. For a more detailed introduction to the model see Matthiessen and Halliday 1992. The grammatical analyses used throughout the volume are based on Halliday 1985a/1993; the discourse and context analyses assumed in Part 2 are drawn from Martin (1992). This overview is organized around the headings: Plane, Metafunction, Stratification, Rank, Realization and Perspective.

Plane — Language and Context

As noted above, SFL treats language and social context as complementary abstractions, related by the important concept of realization (outlined in more detail later). Consider for example the following footnote from Stephen Jay Gould's *Wonderful Life: the Burgess Shale and the nature of history*.

A properly defined group with a single common ancestor is called monophyletic. Taxonomists insist upon monophyly in formal classification. However, many vernacular names do not correspond to well-constituted evolutionary groups because they include creatures with disparate ancestries — 'polyphyletic' groups in technical parlance. For example, folk classifications that include bats among birds, or whales among fishes are polyphyletic. The vernacular term *animal* itself probably denotes a polyphyletic group, since sponges (almost surely) and probably corals and their allies as well, arose separately from unicellular ancestors — while all other animals of our ordinary definitions belong to a third distinct group. The Burgess Shale contains numerous sponges, and probably some members of the coral phylum as well, but this book will treat only the third great group — the coelomates, or animals with a body cavity. The coelomates include all vertebrates and all common invertebrates except sponges, corals, and their allies. Since the coelomates are clearly monophyletic (Hanson, 1977), the subjects of this book form a proper evolutionary group. (Gould, 1989, p. 38)

This footnote is a kind of text — a piece of language. We can write it on the page, read it, it's a thing and we can do things to it. Significantly, it's a semiotic thing, and so the things we do to it are acts of meaning. We call it a text — an instance of language in use. At the same time the text is in some sense about something — in this case the social practice whereby scientists classify living things. Gould is here apprenticing readers into one aspect of this social practice — the convention whereby 'taxonomists insist on monophyly in formal classification'. This social practice can itself be construed semiotically; this is the linguistic reading given to context in social semiotic theory. And since taxonomizing is itself very much a semiotic process (i.e., a reworking of commonsense classifications into un-commonsense ones), this perspective is an especially appropriate one.

In order to model the relationship between language and social context we can use the image of concentric circles as in Figure 2.1. This design is intended to establish one semiotic system (language) as the realization of another more abstract semiotic system (social context). By realization, as suggested earlier, we mean that one system redounds with the other: language construes, is construed by and (over time) reconstrues and is reconstrued by social context. The double-headed arrow in the diagram symbolizes this mutual determination (as opposed to unidirectional notions of cause and effect). Technically we can refer to the distinct levels of abstraction as semiotic 'planes'.

The model allows us to approach Gould's text from two complementary perspectives — the perspective of language (science as text) and the perspective of social context (science as institution). And the model encourages us to be as explicit as possible about the relationship between the two. For example, from the perspective of text, Gould's footnote establishes vertebrate and most invertebrate animals as monophyletic. The key passages are as follows:

a properly defined group with a single common ancestor is called monophyletic

this book will treat only the third great group — the coelomates, or animals with a body cavity

Figure 2.1: Language as the Realization of Social Context

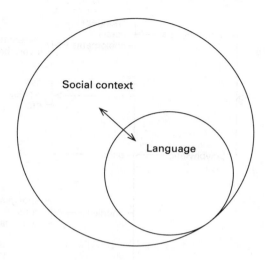

the coelomates are clearly monophyletic

the coelomates include all vertebrates and all common invertebrates except sponges, corals and their allies. (Gould, ibid.)

If we then take the rest of the footnote into account, the text construes the classification of 'grouping principles' in Figure 2.2, exemplifying both the monophyletic and polyphyletic groups. Gould is concerned in his footnote to establish that the subject matter of his book, the coelomates found in the Burgess Shale, form a proper evolutionary group.

Now let's recontextualize this from the perspective of science as institution. To do this we need to broaden our perspective, taking into account as many other scientific texts as we can get our hands on. Working in this way we can start to build up an overall picture of taxonomizing as a social practice, its evolution in the discipline (the special concern of Gould's book) and its interrelationships with other science practices. For example, in the second footnote directly following the footnote just considered, we learn that some flora, unlike fauna, may indeed have disparate ancestries, since hybridization between distinct lineages occurs frequently. Furthermore Gould alerts us to the fact that genes can be transferred laterally by viruses across species boundaries, a process which may have been important in the evolution of some unicellular creatures. And many of us will recognize beyond this that thanks to genetic engineering viruses are no longer the only organisms responsible for transferring genetic material.

What all of this appears to mean is that monophyly is not a principle scientists can depend on in formal classification. They have to take into account whether they're dealing with plants or animals, whether the plants are unicellular or complex and whether genetic transfer across species is involved. As Gould

Figure 2.2:   Construal of Grouping Principles in Text

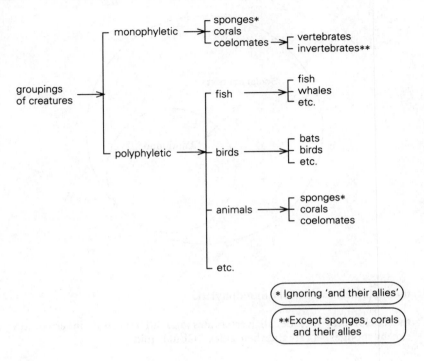

comments, classifications of plants may look more like a network than a conventional diversifying bush, which raises a query about the opening sentence of the footnote considered above: 'Taxonomists insist upon monophyly in formal classification'. Does this mean that a plant network is not a formal classification? Or are we meant to reconstrue Gould's comment on the basis of his second footnote, restricting it to the classification of animals? Gould's book does not resolve this issue, which leaves us in the position of having to find relevant texts that do; and perhaps the easiest way to procure these would be to elicit them from practising scientists by asking them what they do.

Our point here is that a given text provides only a very partial perspective on the social practice of science. This means that in the short term detailed linguistic analysis fleshes out only small parts of the overall picture at a time. Consequently it is only by shunting between language and social context (i.e., between the planes of science as text and science as institution) that we can begin to map out a meaningful interpretation of the discourse of science. As linguists, we treat the institutional perspective as more abstract because it generalizes across a vast range of actual texts (some of which we take as point of departure in the analyses we pursue) and an even larger range of potential verbalizations.

Metafunction — Modes of Meaning

As noted above, the engagement of physical, biological and social resources in our species engenders consciousness, and human consciousness engenders semiosis. As far as we can tell, semiosis engendered in this way, including language, involves three complementary modes of meaning. It is designed to construe biological and physical resources; it is also designed to enact social resources; and in addition it is reflexively designed to organize its own manifestations (i.e., texts) by adapting them to their environment. Let's take another small text from Gould to illustrate this point.

> Step way way back, blur the details, and you may want to read this sequence as a tale of predictable progress: prokaryotes first, then eukaryotes, then multicellular life. But scrutinize the particulars and the comforting story collapses. Why did life remain at stage 1 for two-thirds of its history if complexity offers such benefits? Why did the origin of multicellular life proceed as a short pulse through three radically different faunas, rather than as a slow and continuous rise of complexity? The history of life is endlessly fascinating, endlessly curious, but scarcely the stuff of our usual thoughts and hopes. (Gould, 1989, p. 61)

First, as part of enacting social relations, Gould engages the reader in these distinct tasks (or reading positions) — drawing on the interpersonal resources known as *mood*. These involve: giving the reader information, which he or she is expected to receive; asking the reader for information, which he or she is not in fact expected to possess but will subsequently be all the more grateful to Gould for providing; and asking the reader for services, in this case mental services, which he or she is expected to provide. These three diverse enactments are outlined below, along with their grammatical designations as declarative, interrogative and imperative *mood*:

- giving information (declarative)
  and **you may** want to read this sequence as a tale . . .
  . . . and **the comforting story collapses**.
  The history of life is endlessly fascinating, endlessly curious . . .

- demanding information (interrogative)
  Why did life remain at stage 1 for two-thirds of its history . . .?
  Why did the origin of multicellular life proceed as a short pulse . . .?

- demanding services (imperative)
  Step way way back,
  blur the details,
  But **scrutinize** the particulars

As the same time Gould is concerned to construe physical and biological reality (including consciousness), drawing among other things on the ideational resources referred to as *process type*. This involves: building up a world of action in which physical and biological entities act, by themselves, or on other things; construing a world of semiotic activity in which typically conscious entities negotiate meaning; and constructing a world of relationships among entities — a world in which things can be without doing. These diverse construals of reality are outlined below, along with their grammatical designations as material, verbal/mental and relational processes.

- action (material processes)
  Step way way back,
  blur the details,
  and the comforting story **collapses**.
  if complexity **offers** such benefits
  Why did the origin of multicellular life **proceed** as a short pulse . . .

- Signification (verbal and mental processes)
  and you may want to **read** this sequence as a tale of . . .
  But **scrutinize** the particulars

- being (relational processes)
  Why did life **remain** at stage 1 for two-thirds of its history
  The history of life **is** endlessly fascinating, endlessly curious . . .

Finally, in order to enact social relations and construe reality it is necessary for Gould to mean — to bring into being a semiotic reality alongside the physical, biological and social. In order to adapt this reality to its environment Gould draws on textual resources, including the system for tracking participants referred to as 'identification' in Martin, 1992. It is easy to see for example that the text we've been considering in this section depends on another one, which is explicitly referred to four times: 'the details', 'this sequence', 'the particulars'. 'the comforting story.' The deixis in these nominal groups ('the', 'this', 'the', 'the') begs the question 'which one?'

> Step way way back, blur **the details**, and you may want to read **this sequence** as a tale of predictable progress: prokaryotes first, then eukaryotes, then multicellular life. But scrutinize **the particulars** and **the comforting story** collapses. Why did life remain at stage 1 for two-thirds of its history if complexity offers such benefits? Why did the origin of multicellular life proceed as a short pulse through three radically different faunas, rather than as a slow and continuous rise of complexity? The history of life is endlessly fascinating, endlessly curious, but scarcely the stuff of our usual thoughts and hopes.

Here in fact is the relevant co-text, the immediately preceding paragraph in Gould's book:

> Thus, instead of Darwin's gradual rise to mounting complexity, the 100 million years from Ediacara to Burgess may have witnessed three radically different faunas — the large pancake-flat soft-bodied Ediacara creatures, the tiny cups and caps of the Tommotian, and finally the modern fauna, culminating in the maximal anatomical range of the Burgess. Nearly 2.5 billion years of prokaryotic cells and nothing else — two-thirds of life's history in stasis at the lowest level of recorded

complexity. Another 700 million years of the larger and more intricate eukaryotic cells, but no aggregation to multi-cellular life. Then, in the 100-million-year-wink of a geological eye, three outstandingly different faunas — from Ediacara, to Tommotian, to Burgess. Since then, more than 500 million years of wonderful stories, triumphs and tragedies, but not a single new phylum, or basic anatomical design, added to the Burgess complement. (Gould, 1989, pp. 60–61)

'This sequence' and 'the comforting story' refer generally to the evolutionary narrative Gould constructs; 'the details' and 'the particulars' refer to Gould's specifications. The deixis adapts Gould's subsequent paragraph to his preceding one, logogenetically integrating it with its co-text.

In SFL the three modes of meaning illustrated above are referred to as metafunctions — the interpersonal (social reality), the ideational (physical and biological reality[2]) and the textual (semiotic reality). These modes of meaning organize the communicative planes of both language and social context. Table 2.1 shows the correlations between the language systems just considered, metafunctions and the contextual categories tenor, field and mode. This metafunctional cross-classification of planes is further outlined in Figure 2.3 which projects linguistic metafunctions onto context in terms of the realizational solidarities of interpersonal meaning with tenor, ideational meaning with field and textual meaning with mode.[3] By way of apology, we should clarify that the diagram is not intended to prioritize ideational meaning, nor to imply that field resources are more extensive than those of tenor or mode — our geometry has failed us in this regard!

Stratification — Levels of Semiosis

Levels of Language

Alongside metafunctions, which represent the way language is organized with respect to simultaneous modes of meaning, SFL follows Hjelmslev in viewing language as a stratified system, with one, more 'concrete', stratum (his expression 'plane'[4]) realizing a second more 'abstract' one (his 'content plane'). One way to exemplify these two levels of linguistic abstraction is to consider the difference between an acronym and a technical term. Both acronyms and technical terms are commonly used in scientific discourse to condense information. Acronyms compact information on the expression 'plane'; earlier for example we introduced the acronym SFL in order to avoid having to spell out s-y-s-t-e-m-i-c f-u-n-c-t-i-o-n-a-l l-i-n-g-u-i-s-t-i-c-s each time we mentioned it in this introduction.

Technical terms on the other hand compact information on the content 'plane'; in the footnote considered above for example Gould defined coelomates with the more expanded wording 'animals with a body cavity.' Once defined, the technical term can be assumed in place of the more cumbersome commonsense wording; equally important, the technical term participates in an un-common-sense set of oppositions involving coelomates, sponges, corals and their allies — a precise set of values that can only be construed through the appropriate technical terms. So while abbreviation is a satisfactory characterization of condensation on the expression plane (the acronym), a better metaphor for condensation on the

Table 2.1:   Solidarity across Example Systems, Metafunctions and Contextual Categories

LANGUAGE SYSTEM	METAFUNCTION	CONTEXTUAL CORRELATE
mood	interpersonal	tenor
process type	ideational	field
identification	textual	mode

Figure 2.3:   Metafunctional Solidarity across Planes

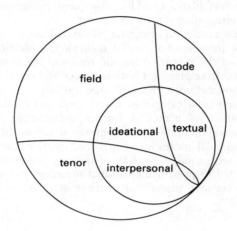

content plane is distillation, since technical terms both compact and change the nature of the more commonsense meanings by which they are defined. The difference between these two types of condensation depends precisely on the principle of stratification, whereby language is doubly articulated as expression form (phonology or graphology), and as content form (lexicogrammar and discourse semantics).

abbreviating expression:
Systemic functional linguistics (hereafter abbreviated to 'SFL')

distilling content:
the coelomates, or animals with a body cavity

Unlike Hjelmslev, SFL takes the further step of explicitly stratifying the content 'plane', setting up a level of lexicogrammar which mediates in realization terms between phonology/graphology and discourse[5] semantics. This stratified content plane provides us with essential resources for interpreting grammatical metaphor. Consider for example the following of Gould's grammatical metaphors, his non-technical condensation of one important aspect of Darwin's theory of evolution: 'Darwin's gradual rise to mounting complexity'. Literally, at the level of lexicogrammar, this is a nominal group; it is grammaticalized as a thing, as outlined in Figure 2.4.

Figure 2.4: *Nominal Construal of One Aspect of Darwin's Theory of Evolution*

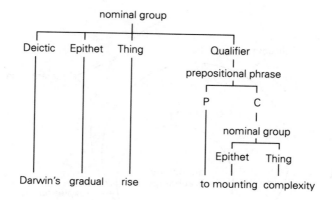

Figure 2.5: *The Tension between Literal and Figurative Readings of Gould's Nominalization of Darwin's Theory*

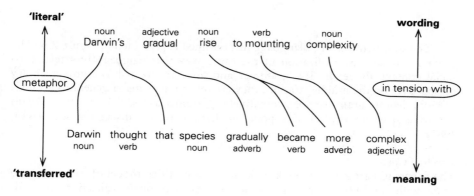

Figuratively however, at the level of semantics, this condensation needs to be unpacked. For one thing, as is typical of heavily nominalized condensations of this kind, it is highly ambiguous. It does not for example refer to Darwin's own physical development during his lifetime, just one among many possible alternatives.[6] Rather, given the field and the relevant co-text, we need to expand this nominalization into a more explicit clause-like form — something along the lines of 'Darwin thought that species gradually became more complex'. The tension between the 'literal' and the 'figurative' reading of this phrase is outlined in Figure 2.5. Given a stratified content plane, this can be thought of as a tension between discourse semantics (the figurative reading) and lexicogrammar (the literal reading). Grammatical metaphors in other words unhinge the two strata, allowing a degree of symbolic play between the two; the play forces us to interpret metaphorical discourse on two levels in order to construe the meaning that is being made in the text.

Figure 2.6:   Stratification: Levels of Abstraction in Language

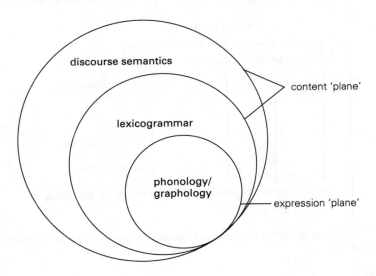

The stratified model of language just outlined is presented in Figure 2.6. The strata are related by realization in the same way that language was related to social context above. In SFL grammar and lexis are treated as complementary perspectives on the middle stratum; they are related in terms of generality — lexis as more delicate grammar, or alternatively, grammar as less delicate lexis. We do not have space to explore this point in detail here; for discussion see Hasan, 1987a.

Levels of Context
The papers in Part 2 of this book make use of a stratified model of context,[7] once again with concentric levels of abstraction related through realization. Accordingly in this section we will present a brief introduction to the model of social context presented in Martin (1992). Taking language as point of departure, this model first contextualizes linguistic choices with respect to the widely used systemic categories of field, mode and tenor. Halliday's 1985b/1989 characterizations have been adapted to introduce these variables below:

FIELD — **the social action**: 'what is actually taking place'
refers to what is happening, to the nature of the social action that is taking place: what is it that the participants are engaged in, in which the language figures as some essential component. 1985b, p. 12 (including the notion of activity sequences oriented to some more global institutional purpose)

TENOR — **the role structure**: 'who is taking part'
refers to who is taking part, to the nature of the participants, their statuses and roles: what kinds of role relationship obtain among the

participants, including permanent and temporary relationships of one kind or another, both the types of speech role that they are taking on in the dialogue and the whole cluster of socially significant relationships in which they are involved. 1985b, p. 12 (including what Halliday 1978, p. 33 refers to as the 'degree of emotional charge' in the relationship)

MODE — **the symbolic organization**: 'What role language is playing' refers to what part language is playing, what is it that the participants are expecting the language to do for them in the situation: 'the symbolic organization of the text, the status that it has, and its function in the context, including the channel (is it spoken or written or some combination of the two?) . . . 1985b, p. 12 (including the degree to which language is part of or constitutive of what is going on and the potential for aural and visual feedback between interlocutors)

In Martin's model, these three variables taken together comprise the level of register.[8] Choices within these variables put at risk the linguistic choices which construe context of situation. To exemplify this framework, let's briefly reconsider Gould's synoptic history of evolution. How do we know what this small text is on about? From the perspective of field, we recognize the text as enacting science — paleontology to be exact. This is construed for us by its technical terms (e.g., 'fauna', 'prokaryotic cells', 'geological', 'phylum'), the orientation to explicit measurement (e.g., '100 million years', '2.5 billion years') and its indexical names ('Darwin', 'Burgess') — alongside other features which we will not unpack here. The features just reviewed are highlighted in bold face below:

FIELD (highlighting technicality):

Thus, instead of **Darwin**'s gradual rise to mounting complexity, the **100 million years** from **Ediacara** to **Burgess** may have witnessed three radically different **faunas** — the large pancake-flat **soft-bodied Ediacara creatures**, the tiny cups and caps of the **Tommotian**, and finally the **modern fauna**, culminating in the maximal **anatomical range** of the **Burgess**. Nearly **2.5 billion years** of **prokaryotic cells** and nothing else — two-thirds of life's history in stasis at the lowest level of recorded complexity. Another **700 million years** of the larger and more intricate **eukaryotic cells**, but no aggregation to **multi-cellular life**. Then, in the **100-million-year**-wink of a **geological** eye, three outstandingly different **faunas** — from **Ediacara**, to **Tommotian**, to **Burgess**. Since then, more than **500 million years** of wonderful stories, triumphs and tragedies, but not a single new **phylum**, or basic **anatomical design**, added to the **Burgess** complement. (Gould, 1989/1991, pp. 60–61)

At the same time, from the perspective of mode, we recognize the text as abstract — as abstract writing to be exact. The main feature of the text that clues us into the reflective nature of this text is its nominal style. The whole text contains only three verbs ('may have witnessed', 'culminating' and 'added'). Two of these are in the first sentence and one in the last; the verb in the last sentence is embedded, qualifying . . . 'new phylum, or basic anatomical design'; and one of the verbs, 'culminating', is actually a metaphorical realization of a logical temporal relation — it does not construe action. Gould's second, third and fourth sentences contain no verbs at all! By nominalizing qualities (e.g., 'complexity') and processes (e.g., 'aggregation', 'triumphs') Gould renders 3.8 billon years of life on our planet as an abstract succession of things. The key features of this nominal style are highlighted below (nominalizations in bold face, verbs in italics).

MODE (highlighting abstraction)

> Thus, instead of Darwin's gradual **rise** to mounting **complexity**, the 100 million years from Ediacara to Burgess *may have witnessed* three radically different faunas — the large pancake-flat soft-bodied Ediacara creatures, the tiny cups and caps of the Tommotian, and finally the modern fauna, *culminating* in the maximal anatomical **range** of the Burgess. Nearly 2.5 billion years of prokaryotic cells and nothing else — two-thirds of life's **history** in **stasis** at the lowest **level** of recorded **complexity**. Another 700 million years of the larger and more intricate eukaryotic cells, but no **aggregation** to multi-cellular **life**. Then, in the 100-million-year-**wink** of a geological eye, three outstandingly different faunas — from Ediacara, to Tommotian, to Burgess. Since then, more than 500 million years of wonderful **stories**, **triumphs** and **tragedies**, but not a single new phylum, or basic anatomical **design**, *added* to the Burgess **complement**. (Gould, 1989/1991, pp. 60–61)

Strikingly, Gould assumes this nominal posture without in any way taking the excitement out of the puzzle he wishes to pose. From the point of view of tenor, the text is authoritative; Gould is giving, not negotiating information. And the text is relatively exclusive; only biologists and readers who have been apprenticed by Gould's book into the field will be able to follow what is going on. But no-one could miss the impact Gould constructs by amplifying his thesis to the point where Darwin's antithesis lacks credibility. It's not logic Gould is deploying here, but rhetoric; the argument hangs on the emotional impact Gould construes by grading quantity from low to high ('lowest', 'more', 'maximal'); denying exception ('nothing else', 'no aggregation', 'not a single new phylum'), constructing goals ('finally', 'culminating'), intensifying difference ('radically different', 'outstandingly different') and maximizing evaluation ('wonderful stories', 'triumphs and tragedies'). Gould's appraisal of the argument is outlined below, followed by a reading of the text highlighting the subjective intrusions he positions his readers to share.

Darwin's antithesis	*Gould's thesis*
MODERATE COMPLEXITY **gradual** rise to **mounting** complexity	AMPLIFIED COMPLEXITY **lowest** level of recorded complexity, **larger** and **more** intricate eukaryotic cells, **maximal** anatomical range of the Burgess

DENIAL OF EXCEPTION
nothing else, **no** aggregation to multicellular
life, **not a single** new phylum . . .

TELOS
finally, culminating

INTENSIFICATION
radically different, **outstandingly** different

EVALUATION
wonderful stories, **triumphs** and **tragedies**

TENOR (highlighting appraisal)

Thus, instead of Darwin's **gradual** rise to **mounting** complexity, the
100 million years from Ediacara to Burgess may have witnessed three
radically different faunas — the large pancake-flat soft-bodied Ediacara
creatures, the tiny cups and caps of the Tommotian, and **finally** the
modern fauna, **culminating** in the **maximal** anatomical range of the
Burgess. Nearly 2.5 billion years of prokaryotic cells and **nothing else**
— two-thirds of life's history in stasis at the **lowest level** of recorded
complexity. Another 700 million years of the **larger** and **more intricate**
eukaryotic cells, but no **aggregation** to multi-cellular life. Then, in the
100-million-year-wink of a geological eye, three **outstandingly different** faunas — from Ediacara, to Tommotian, to Burgess. Since then,
more than 500 million years of **wonderful stories, triumphs and
tragedies**, but **not a single** new phylum, or basic anatomical design,
added to the Burgess complement. (Gould, 1989/1991, pp. 60–61)

Taken together, this syndrome of technicality, abstraction and appraisal
constructs a familiar register, that of popular science, for which the author,
Stephen Jay Gould, is renowned. The papers in Part 2 of this volume draw on a
more elaborated model of context which evolved to complement the metafunctionally diversified picture of register illustrated above. This model recon-
textualizes register (field, mode and tenor) with the more abstract level of genre.
Genre theory provides a more holistic perspective on contextual analysis, which
interfaces with important traditions of research in literary theory, cultural studies
and mass communication. Perhaps the most engaging interface is provided through
the work of Bakhtin, who writing in 1952–1953 anticipates systemic interpretations of context in striking ways:

All the diverse areas of human activity involve the use of language.
Quite understandably, the nature of forms of this use are just as diverse
as are the areas of human activity . . . Language is realized in the form of
individual concrete texts (oral and written) by participants in the various
areas of human activity. The texts reflect the specific conditions and
goals of each such area not only through their content (thematic) and
linguistic style, that is the selection of the lexical, phraseological, and
grammatical resources of the language, but above all through their

compositional structure. All three of these aspects — thematic content, style, and compositional structure — are inseparably linked to the whole of the text and are equally determined by the specific nature of the particular sphere of communication. Each separate text is individual, of course, but each sphere in which language is used develops its own relatively stable types of these texts. These we may call 'speech genres'. (Bakhtin, 1986, p. 60, writing in 1952–1953; the term 'text' has been substituted for 'utterance' throughout)

Genre theory, as developed in SFL, had been particularly concerned with texts as staged goal-oriented social processes which integrate field, mode and tenor choices in predictable ways. Seen in these terms, Gould's synoptic history of evolution unfolds as a kind of narrative. Life begins (prokaryotic cells), develops (eukaryotic cells), diversifies (Ediacara and Tommotian creatures), explodes into the modern fauna of the Burgess Shale and then subsequently subsides through large-scale attrition until we arrive at the ever diminishing number of extant species co-habiting the planet today. To make his point Gould rewrites Darwin's model of evolution as a puzzling story, whose climax (the Burgess Shale) is clear, but whose interpretation is a matter of considerable debate. As would be predicted, given the narrative genre, Gould scaffolds his text very heavily with respect to sequence in time (it is this strongly foregrounded temporal scaffolding which enables Gould to construct a story out of nominal groups). The temporal scaffolding in Gould's history is foregrounded below:

GENRE (highlighting time)

> Thus, instead of Darwin's gradual rise to mounting complexity, **the 100 million years from Ediacara to Burgess** may have witnessed three radically different faunas — the large pancake-flat soft-bodied Ediacara creatures, the tiny cups and caps of the Tommotian, and **finally** the modern fauna, **culminating** in the maximal anatomical range of the Burgess. **Nearly 2.5 billion years** of prokaryotic cells and nothing else — two-thirds of life's history in stasis at the lowest level of recorded complexity. **Another 700 million years** of the larger and more intricate eukaryotic cells, but no aggregation to multi-cellular life. **Then, in the 100-million-year-wink of a geological eye**, three outstandingly different faunas — **from** Ediacara, **to** Tommotian, **to** Burgess. **Since then, more than 500 million years** of wonderful stories, triumphs and tragedies, but not a single new phylum, or basic anatomical design, added to the Burgess complement. (Gould, 1989/1991, pp. 60–61)

We will not take time to weigh up the pros and cons[9] of stratified as opposed to unstratified models of context here. Practically speaking, it can simply be noted at this point that the stratified model has proven very effective in Australian educational linguistics as far as giving teachers an accessible handle on contextual considerations is concerned. Numerous references to these initiatives are given in the papers in Part 2. Theoretically, the model has opened up a very productive dialogue in Australian social semiotic theory, involving systemic linguistics, critical theory, feminist theory and post-structuralism (interesting records of which are found in Threadgold *et al.* 1986, Giblett and O'Carroll, 1990 and Christie, 1991). It is too early to predict the directions in which systemic models of social context will resolve.

Finally it should be noted that the dialogue of theory and applications just reviewed has led to a recontextualization of genre by a more abstract level of ideology. Ideology in this model focuses on social subjectivity — with the ways in which discourses of class, gender, ethnicity and generation position speakers as far as the meaning potential of their culture is concerned. It might be argued for example that Gould's popular science constitutes a middle-class, male, white, adult discourse which is certain to exclude large numbers of readers: he construes research as competition, applauds discovery as victory and dreams of conferring a Nobel prize on his heroes; he shares pornographic mnemonics for the geological time scale with readers; he 'deconstructs' eight figures (1.5–1.12) as falsely portraying the progressive evolution of humans from apes, without addressing their racist or sexist discourses (all but one exclude women; the figure including women makes a 'joke' by showing men as evolving, while women remain the same, on all fours, scrubbing the floor;[10] in some figures the final step is more plausibly caucasian looking than the penultimate one; one figure makes a 'joke' by having a terrorist with Middle Eastern or Indian head-dress parachuting into place on the lineage far back among the apes[11]); and Gould's generally high level of grammatical metaphor automatically excludes all but the most literate, university-trained readers.

Gould may be the best in his trade; but his trade is itself naturalized by dominant discourses with which he does not contend. Gould may be popularizing science, but we need to ask who he is popularizing science for. In what sense is an audience comprised of white, male, middle-class, lay intellectuals a popular one? Just how relevant is popular science writing of this kind to the apprenticeship of female, migrant or working-class children into science in schools? The recontexualization of genre as ideology in the model has been designed to make way for deconstructions of social subjectivity, coding orientation and access to meaning potential along the lines of inquiry opened up by queries such as these.

The stratified model of context reviewed here is outlined in Figure 2.7. In summary, social context is realized by language; at the level of social context, ideology is realized by genre, which is in turn realized by register. In Hjelmslev's terms, ideology, genre and register are connotative semiotics, because they make use of another semiotic system (i.e., language) as their expression form; language itself is a denotative semiotic because it has its own expression form (i.e., phonology or graphology). We should perhaps stress here that language is just one of the denotative semiotics through which our culture construes meaning; others, including music, dance, kinesics, proxemics, architecture, images, figures and diagrams, film, video, photography and so on are critically involved. The extent to which the connotative semiotics construed by these denotative ones are homologous with register, genre and ideology as introduced above remains a very open question at this time.

Rank — Constituency within Strata

In order to take the next step in our introduction to the model, we need to subdivide the ideational metafunction into its logical and experiential sub-components.

Figure 2.7:   Language in Relation to its Connotative Semiotics: Ideology, Genre and Register

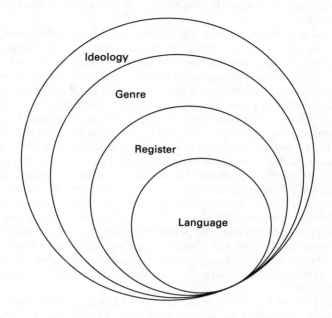

Overall, ideational meaning construes reality by segmenting it; the logical metafunction is concerned with part–part relations across segments, whereas the experiential metafunction is concerned with part–whole relations between segments and the whole they compose. This complementarity of part–part and part–whole relations is important for understanding the way in which grammatical metaphor polarizes the construal of reality in spoken and written modes.

Take for example the following sentence in which Gould uses a verbal process ('he insisted') to project what Darwin argued ('that any complex Cambrian creature must have arisen . . .'):

> He [Darwin] insisted that any complex Cambrian creature must have arisen from a lengthy series of Precambrian ancestors with the same basic anatomy. (Gould, 1989/1991, p. 272)

The relationship between the projecting and projected clause is handled in SFL via the logical metafunction, which construes a part–part relation between the clauses, with the second dependent on the first (this contrasts with projecting Darwin's actual wording[12], instead of the gist of what he argued, as in 'Any complex Cambrian creature must have arisen from a lengthy series of Precambrian ancestors', he insisted):

α   ¨β

He insisted   that any complex Cambrian creature must have arisen from
                  a lengthy series of Precambrian ancestors with the same
                  basic anatomy.

One important feature of these part–part relations is that they are indefinitely recursive; for example:

Darwin insisted that any complex Cambrian creature must have arisen from a lengthy series of Precambrian ancestors . . .

Gould wrote that Darwin insisted that any complex Cambrian creature must have arisen from a lengthy series of Precambrian ancestors . . .

Halliday and Martin reported that Gould wrote that Darwin insisted that any complex Cambrian creature must have arisen from a lengthy series of Precambrian ancestors . . .

Earlier in the book, Gould construes roughly the same meaning in a third way, this time not as a dependent or independent clause, but as a nominal group (in bold face below):

Yet the peculiar character of this evidence has not matched Darwin's prediction of **a continuous rise in complexity towards Cambrian life**. . . (Gould, 1989/1991, p. 57)

Rendered in this form the meaning is clearly part of something else and is handled in SFL via the experiential metafunction, which construes a part–whole relation between 'a continuous rise in complexity towards Cambrian life' and the rest of its prepositional phrase, which is in turn part of a nominal group in which it qualifies the nature of Darwin's prediction, which in turn functions as part of an identifying relational clause in which it gives value to the character of the evidence provided by the rich Precambrian record discovered in the past thirty years. This compositional construal of reality is outlined in Figure 2.8, using a constituency diagram.

Our general point here is that nominalization downgrades the grammatical status of meanings, so that what might be construed as a combination of inter-dependent clauses in the spoken mode is reconstrued as edifice of words and phrases in writing. The meaning comes to function, in other words, at a lower rank in the grammar — at the ranks of group/phrase and word, instead of at the rank of clause. This has tremendous implications for the texture of the discourse unfolding in this way. For one thing, it is less negotiable, since you can argue with a clause but you can't argue with a nominal group. Gould's construal of Darwin's prediction of 'a continuous rise in complexity' as a thing means that it is taken for granted that Darwin did in fact predict such a development; it cannot easily be challenged. At the same time, nominalization of this kind opens up a vast potential for distributing and redistributing information in the clause. In nominal form Darwin's prediction can be placed first in the clause as Theme, or last as New; it can be made both Theme and New; it can be exclusively identified as News or Theme, and so on; for example:

Yet the peculiar character of this evidence has not matched **Darwin's prediction of a continuous rise in complexity towards Cambrian life**

Figure 2.8: Constituency Representation for Part–Whole Relations

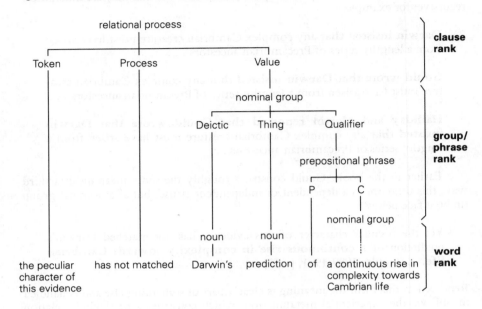

Yet **Darwin's prediction of a continuous rise in complexity towards Cambrian life** has not matched the peculiar character of this evidence

Yet it is **Darwin's prediction of a continuous rise in complexity towards Cambrian life** which has not matched the peculiar character of this evidence

What has not matched the peculiar character of this evidence is **Darwin's prediction of a continuous rise in complexity towards Cambrian life**

What **Darwin's prediction of a continuous rise in complexity towards Cambrian life** has not matched is the peculiar character of this evidence

And at the same time Darwin's prediction can be made to function as a participant in a vast array of relational clause types which can be deployed to describe, to classify, to exemplify, to identify, to decompose, to order, to interface, to contrast, to explain, to prove and so on. Scientific English has in fact been expanding the meaning potential in this area of the grammar by leaps and bounds for over three hundred years to the point where hundreds of verbs are routinely deployed which relate these nominalizations to one another. Here a few tokens of this semogenetic explosion:

Darwin's prediction of a continuous rise in complexity towards Cambrian life **ended up** incorrect.

Darwin's prediction of a continuous rise in complexity towards Cambrian life **looked** a mistake.

Darwin's prediction of a continuous rise in complexity towards Cambrian life **exemplifies** what we're criticizing here.

Darwin's prediction of a continuous rise in complexity towards Cambrian life **represents** one of his errors.

Darwin's prediction of a continuous rise in complexity towards Cambrian life **constitutes** just a part of his theory.

Darwin's prediction of a continuous rise in complexity towards Cambrian life **follows on from** his basic theory.

Darwin's prediction of a continuous rise in complexity towards Cambrian life **correlates** precisely with his principle of natural selection.

Darwin's prediction of a continuous rise in complexity towards Cambrian life **differs from** predictions we might make based on current evidence.

Darwin's prediction of a continuous rise in complexity towards Cambrian life **led to** his preoccupation with the fossil record.

Darwin's prediction of a continuous rise in complexity towards Cambrian life **proves** the point we alluded to earlier.

In short then, a radical expansion in the areas of textual and experiential meaning potential is bought at the interpersonal price of decreasing negotiability, since down-ranked meanings are relatively difficult to challenge. This problem is something that needs to be seriously addressed as science discourse moves into a century in which it has to negotiate in new ways with both discursive and non-discursive resources if semiotic, or even biological resources on the planet are to be encouraged to survive.

Realization

The concept of realization has already been introduced, by way of characterizing the relationship between levels of abstraction, including planes and strata, in SFL. Following Lemke (1984) realization is most effectively interpreted as a chain of metaredundancy — a redundancy on one level is redundant with part of a redundancy on another level, which is in turn redundant with part of a redundancy on a further level and so on. For example, the phonological redundancy[13] /sistiymik/, redounds at the level of lexicogrammar with the word 'systemic'; this word

Figure 2.9: A Partial System Network for 'Being' Clauses

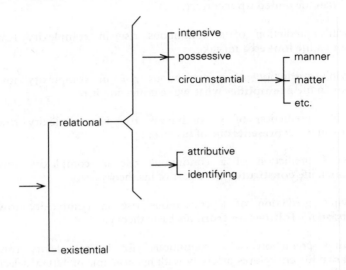

enters into syntagmatic redundancies at the level of lexicogrammar such as 'systemic functional linguistics' which redound at the level of discourse semantics with a participant denotable as 'systemics'; this participant enters into identification chains at the level of discourse semantics such as 'systemics-it-it-the theory-other theories', which in turn redound with the taxonomies of theories of language at the level of field; and so on. Throughout these examples redundancy can be interpreted as non-random co-occurrence — a recognizable pattern in other words. Metaredundancy theory has the advantage of being directional as far as degrees of abstraction are concerned, but non-directional as far as cause and effect are concerned. It also has the advantage of being probabilistic rather than categorical, with categorical realization interpretable simply as 100 per cent probability of co-occurrence (the exception rather than the rule); this makes room for one level of abstraction to reconstrue another and for the system to evolve.

The term realization is also commonly used in another way in SFL to refer to the instantiation of system in process. In order to explore this sense of realization we need to keep in mind that SFL represents language as system as a complex network of relationships. These relationships are modelled as interconnecting networks of systems. A partial network for 'being' clauses is presented in Figure 2.9. This network is located at clause rank, in the lexicogrammar, in the experiential metafunction, in language.

Following Halliday 1985a, the network first distinguishes between existential processes and relational ones; existential clauses have only one obligatory participant (the Existent), whereas other 'being' clauses have two. In order to show *mood*, existential clauses usually have a Subject realized by 'there'; for example, see Figure 2.10.

Relational processes, as their name implies, have two obligatory participants which are related to each other in one of two general ways — through attribution or through identification. The basic grammatical distinction between the two is

Figure 2.10:   *Experiential and Interpersonal Analysis*

	Process	Existent	Circumstance	
	Process	Existent	Circumstance	**experiential analysis**
Subject	Finite	Complement	Adjunct	**interpersonal analysis**
There	*were*	*numerous phyla*	*in the Burgess Shale*	

Figure 2.11:   *The Attributive Clause*

	attributive	identifying
'active'	The coelomates are monophyletic.	The term animal *denotes* a polyphyletic group.
'passive'		A polyphyletic group *is denoted by* the term animal.

Figure 2.12:   *Relational-Clauses Paradigm*

	ATTRIBUTIVE	IDENTIFYING
INTENSIVE	the coelomates **are** monophyletic	the term animal **denotes** a polyphyletic group
POSSESSIVE	other animals **belong** to a third group	the coelomates **include** all vertebrates
CIRCUMSTANTIAL	the Paleozoic era **lasted** approximately 345 million years	vernacular names do not **correspond** to well-constituted evolutionary groups

that identifying clauses have active and passive pairs, whereas attributive clauses have no passives; see Figure 2.11.

At the same time, relational clauses can be intensive, possessive or circumstantial. The attributive/identifying and the intensive/possessive/circumstantial systems are simultaneous systems in the network in Figure 2.9 because they cross-classify clauses. They are enclosed by a right-facing brace to signal this cross-classification. The relevant paradigm is as described in Figure 2.12.

The network in Figure 2.9 makes available further choices for circumstantial processes. These choices specify the kind of circumstantial meanings available in both the attributive and identifying modes. In order to simplify the presentation, only the 'manner' and 'matter' options which will be illustrated below have been included.

So much for system — for the meaning potential available in this area of the grammar. What about its instantiation? Gould's footnote on grouping principles can be drawn on again here to illustrate this point, since all of its clauses except two are relational clauses. This foregrounding of relational processes is redundant with the text's genre — a taxonomizing report; and with its field — in particular the taxonomy of grouping principles it construes. The relational processes in the text are classified, item by item, below (along with the single material and the single verbal process).

CIRCUMSTANTIAL/IDENTIFYING

... many vernacular names do not **correspond** to well-constituted evolutionary groups but this book will **treat** only the third great group ...

POSSESSIVE/ATTRIBUTIVE

while all other animals of our ordinary definitions **belong** to a third distinct group.

POSSESSIVE/IDENTIFYING

because they **include** creatures with disparate ancestries ...
that **include** bats among birds, or whales among fishes[14],
The Burgess Shale **contains** numerous sponges ...
The coelomates **include** all vertebrates ...
the subjects of this book **form** a proper evolutionary group.

INTENSIVE/ATTRIBUTIVE

For example, folk classifications ... **are** polyphyletic.
Since the coelomates **are** clearly monophyletic (Hanson, 1977),

INTENSIVE/IDENTIFYING

A properly defined group with a single common ancestor **is called** monophyletic.
The vernacular term 'animal' itself probably **denotes** a polyphyletic group,

MATERIAL[15]

since sponges ... and probable corals ..., **arose** separately from unicellular ancestors —

VERBAL

Taxonomists **insist** upon monophyly in formal classification.

Of special note is the large number of possessive processes used by Gould to build up the taxonomy outlined above as Figure 1.3. This metaredundancy is in one sense counterintuitive, why does Gould make use of so many part–whole relations to construe relations between classes and subclasses? In fact what we have here is an example of process working against system. In presenting taxonomies scientists generally work from more general classes to more specific ones; their texts regularly unfold in this way. In the grammar, on the other hand, the major clause resource for construing class/subclass relations is the intensive/attributive clause; and this clause type works in precisely the opposite direction — from subclasses to classes:

sub-class		class
folk classifications	are	polyphyletic
the coelomates	are clearly	monophyletic

And as noted above, this clause type is not reversible (it has no passive). This forces scientists to move laterally in the grammar and draw on closely related resources that work in the right direction — typically identifying possessive clauses (which are reversible) and attributive possessive clauses (which work from whole to part):

class		subclass
they	include	creatures with disparate ancestries
the coelomates	include	all vertebrates

As far as we are aware, for scientists this grammatical manoeuvre has not reconstrued classification taxonomies as compositional ones. The potential confusion is certainly there for young apprentices however, underlining the importance of making room for two distinct types of realization in the model: realization across levels of abstraction (metaredundancy) and realization between system and process (instantiation). For apprentices, in other words, there are two types of challenge in interpreting taxonomizing reports. As far as meta-redundancy is concerned, grammatical resources that are usually used to construe part–whole relations are used to construe class/subclass ones. As far as in-stantiation is concerned, a large number of relational clause resources are drawn on which have evolved in written discourse and which may be quite unfamiliar to apprentices who have spent most of their life in an oral culture. Note for example the following struggle with identification, projection and nominalization in the opening sentence of a Year 11 exposition:

The issue shown here in this picture shows a protest against the cutting down of the forest.

Both types of realization (instantiation and metaredundancy) have proven very important in promoting dialogue between the systemic model of text and context and the concept of intertextuality, deriving from the work of Bakhtin. It is thus important to underline here the fact that instantiation automatically relates process to system, and thus more or less directly to all other texts which have or could have instantiated that system. The clause 'since the coelomates are clearly monophyletic' for example would be analysed experientially as a Carrier Process Attribute structure:

Carrier	Process	Attribute
the coelomates	*are*	*monophyletic*

From this analysis we know that the clause instantiates a particular set of features from the relational clause network: it's an intensive attributive relational clause. In this respect it instantiates the same selections as does 'folk classifications are polyphyletic'; it differs with respect to one selection both from 'all other animals . . . belong to a third distinct group', which is attributive but possessive, and from 'the vernacular term animal probably denotes a polyphyletic group', which is intensive but identifying; it differs in more significant ways from an existential clause like 'there were numerous phyla in the Burgess Shale'; and it differs more significantly still from the verbal process 'Taxonomists insist upon monophyly in formal classification'. All of these intertextual relations follow on precisely from the system/process relations formalized in SFL at all levels of the model.

Metaredundancy (system-to-system realization across strata and planes) is also crucial. Through this concept texts are related not only to the systems they directly instantiate, but to all other systems in the model as well. Gould's footnote for example redounds with the register variable field; as noted above, it construes the taxonomy of grouping principles assembled in Figure 2.2. This taxonomy in turn redounds with all the ensuing mentions of monophyly and polyphyly in Gould's book, and in all the texts already written or available to be written in the field. Metaredundancy, in other words, relates Gould's footnote to all of the other discussions of and practices of classification in the field, existing or imaginable. As with instantiation, it does this automatically through the formalization of system–system relations in SFL across all levels of the model.

In passing it should be noted that a modular system/process theory of the kind just outlined provides a far more powerful theory of intertextuality than that deriving from Bakhtin. For one thing it makes explicit degrees of intertextual proximity, allowing for precise statements of just how, and just how directly, texts are related. In addition it models language and context as an immanent semiotic potential; in doing so it makes room for relations among texts at hand (one with another) and as well among: texts at hand; texts which have been lost forever from the phylogenetic record (i.e., the countless unrecorded spoken texts and written texts with no 'fossil' record); and texts which may be instantiated in the recognizable future. It is only by modelling semiosis as an immanent intersubjective resource that a robust account of intertextual relations of this kind can be achieved. Recently fashionable approaches to intertextuality which reject in principle the possibility of system can be seen in these terms to provide relatively unconstrained and impoverished accounts of text/text relations. We have tried in our own work to offer a richer and more substantive interpretation.

Perspective: Synoptic and Dynamic Orientations

In closing our introduction to the model we would like to focus briefly on one aspect of intertextuality which was not given prominence above, the notion of co-textual relations (or cohesion), in order to exemplify synoptic and dynamic perspectives on meaning. One of language's three main modes of meaning, the textual metafunction, is very much preoccupied with adapting text to co-text; and as noted earlier, one stratum of language, discourse semantics, proposes text as the basic unit of meaning construing social context. It is on the basis of these texturing modules that we construe cohesive relations between parts of a text (and between one text and relevant parts of its instantiated intertexts). By analysing cohesive relations among the parts of a text as it unfolds, we can obtain a picture of the dynamics of texture — the way in which instantiations at one point in a text put at risk those that ensue, or, to put this the other way round, the way in which instantiations at one point in a text were conditioned by earlier instantiations. Consider for example the following pieces of highly cohesive co-text from Gould, in the order in which they appear in his book:

a rich Precambrian record of precursors for the first complex animals [57]

before the lowest Silurian stratum was deposited, long periods elapsed . . . and that during these vast . . . periods of time, **the world swarmed with living creatures** (Darwin 1859, p. 307). [57]

a continuous **rise in complexity** towards Cambrian life [57]

. . . gradual **rise to mounting complexity** [59]

any complex Cambrian creature must have arisen from **a lengthy series of Precambrian ancestors with the same basic anatomy.** [272]

As the phrases highlighted in bold face reveal, Gould is concerned in these passages to construct a picture of Precambrian life in which evolution was well under way. Reflecting this general orientation, the co-text is strongly integrated through lexical cohesion (after Halliday and Hasan, 1976, Hasan, 1985/1989, Martin, 1992). Some of the more important of these ties are summarized and classified below; items in the strings appear in the order in which they unfold in Gould's text:

precursors-animals-living creatures-creature-ancestors	**['synonymy'**[16]**]**
Precambrian-Silurian-Cambrian-Cambrian-Precambrian	**[co-hyponymy]**
long periods-vast periods of time	**['synonymy']**
complex-complexity-complexity-complex	**[repetition]**
rise-rise-arisen	**[repetition]**
rise . . . mounting	**['synonymy']**
continuous-gradual	**['synonymy']**

By reading this lexical integration back against the grammar of the clauses instantiating these cohesive strings, we can construe, albeit informally, a dynamic perspective on Gould's text. As far as the Precambrian record is concerned, Gould begins by previewing Darwin's own construal of the period, which he then exemplifies;[17] subsequently Gould consolidates Darwin's prediction in nominalized form, which he next compacts by removing projection and circumstantiation (i.e., getting rid of 'prediction and towards Cambrian life'); much later in the book, possibly because a compacted consolidation of this kind would be too elliptical to be effectively interpreted, Gould reviews Darwin's position in a far less nominalized form. This texture of shifting abstraction is outlined below:

DYNAMIC PERSPECTIVE — shifting abstraction:

PREVIEW
Darwin recognized that his theory required a rich Precambrian record of precursors for the first complex animals [57]

EXEMPLIFY
If my theory be true, it is indisputable that before the lowest Silurian stratum was deposited, long periods elapsed . . . and that during these vast, yet quite unknown periods of time, the world swarmed with living creatures (Darwin 1859, p. 307). [57]

CONSOLIDATE
Darwin's prediction of a continuous rise in complexity towards Cambrian life [57]

COMPACT
Darwin's gradual rise to mounting complexity [59]

REVIEW
He [Darwin] insisted that any complex Cambrian creature must have arisen from a lengthy series of Precambrian ancestors *with the same basic anatomy*. [272]

Beyond this dynamic reading of the co-text, as linguists we might also be interested in standing back and considering the way in which Gould construes Darwin's picture of Precambrian life from the perspective of mode. From this perspective we are not so much concerned with how Gould moves from more to less abstract construals and back again as his text unfolds as in the degrees of abstraction used by Gould to construct his meaning. In order to explore this in detail it is useful to operate on the text as an analytic artefact — to reorganize the co-text into a chain of increasing abstraction, from most 'spoken' to most 'written' form.

From this synoptic, after the fact, perspective, Darwin's own construal of Precambrian life is the least nominalized rendering — a world **swarming** with living creatures for very long periods of time; this 'bee-hive' is alternatively segmented by Gould into a lengthy **series** of ancestors; the bee-hive is also rendered semiotically as a rich **record** of precursors (referring not to the animals but to the signs of their existence); more abstract again is Gould's nominalization of Darwin's projecting process (the noun 'prediction' in place of the verbs 'insisted' and 'recognized'), with Darwin's projection (i.e., 'a continous rise in complexity') in post-modifying position — and note that in this kind of construal the animals in question are not explicitly mentioned, but have to be inferred from the nominalization 'complexity' (i.e., Complex what? — Oh yeah, animals); maximum abstraction for this meaning is achieved by eliding Darwin's projecting process completely, leaving Darwin as directly pre-modifying his prediction — as if he somehow owned his meaning or it was to be construed as part of him. By this final stage Darwin has become less a person than a text (see Matthiessen, 1992); Gould has reconstrued the scientist as the body of meanings he made. A synoptic overview of these degrees of abstraction, from least to most abstract, is presented below (significant points of contrast are in bold type):

SYNOPTIC PERSPECTIVE — degrees of abstraction

ACTIVITY AS FACT
If my theory be true, it is indisputable that before the lowest Silurian stratum was deposited, long periods elapsed . . . and that **during these**

vast, yet quite unknown periods of time, the world swarmed with living creatures (Darwin 1859, p. 307). [57]

PROJECTION AS HISTORICAL SEGMENTATION
He [Darwin] insisted that any complex Cambrian creature must have arisen from **a lengthy series of Precambrian ancestors** *with the same basic anatomy.* [272]

PROJECTION AS SEMIOTIC TEXTUALIZATION
Darwin recognized that his theory required **a rich Precambrian record of precursors** for the first complex animals [57]

PROJECTING/PROJECTED ABSTRACTION
Darwin's **prediction of a continuous rise in complexity** towards Cambrian life [57]

DE-PROJECTED ABSTRACTION
Darwin's gradual rise to mounting complexity [59]

Resources for nominalization have provided us with a simple illustration of the need for both dynamic and synoptic perspectives when approaching the language of science. The dynamic lens treats text as an unfolding and contingent process. Among other things this perspective is particularly relevant to pedagogy — to the ongoing apprenticeship of students into science discourse, where what happens next has very much to be reasoned about in terms of what has gone before and what might follow. The synoptic lens on the other hand treats text as a cultural artefact — as an object to be taken apart, interpreted, reassembled and observed. Among other things this perspective is particularly relevant to curriculum, where what is achieved overall is at issue, and has to be reasoned about as components of the larger picture. This text as process/text as object complementarity is fundamental to the deconstruction of meaning in SFL.

Turning from text to system, the same complementarity informs our interpretation of semiotic systems as dynamic open systems — as systems which evolve, non-catastrophically, in ways which makes it possible for our species to survive. The dynamic perspective informs our conception of semogenesis in the life of the individual, or ontogenesis, and in the life of the species, or phylogenesis (semogenesis in text, or logogenesis was reviewed above). Looking dynamically at system means looking at semiotic evolution, with evolution interpretable as expending meaning potential. At the same time the synoptic perspective informs our conception of metastability as it inheres in the semiotic systems we construe as phonology, grammar, discourse semantics, register, genre and ideology. Looking synoptically at system means taking synchronic snapshots of the meaning potential we construe as immanent,[18] from the point of view of our reading position in our time. Lacking a dynamic perspective, it would be impossible for us to appreciate how system can change; at the same time, lacking a synoptic perspective it would be impossible to appreciate the social fact of communities founded on the relatively unconscious intersubjective resources we construe as semiotic systems. Taken together, the two perspectives give us a model of semiotic inertia (i.e., naturalization) and semiotic change (i.e., evolution).

Notes

1   Readers familiar with systemic linguistics will be aware of some differences between Halliday's and Martin's perspectives. Except for a few footnotes in relevant contexts these will be ignored in the discussion here as they do not bear critically on the interpretation of the papers in this volume.

2   Ideational meaning in fact incorporates a theory of both social and semiotic reality, which we must however pass over here (see Matthiessen, 1992 for discussion).

3   In general Halliday has tended to follow Hjelsmlev in treating the line between content form and expression form as (proto) typically arbitrary; hence phonology is not given a metafunctional reading (and prosodic systems of intonation are interpreted lexicogrammatically). The alternative reading introduced diagrammatically here draws on Firth, with all levels of language, including expression form, interpreted as meaning making (and thus organized by metafunction).

4   SFL uses the term stratum for Hjelmslev's plane; Martin, 1992 redeploys the term plane to describe the difference between a denotative and its connotative semiotics — the levels of language and social context for example.

5   Martin uses the term discourse semantics as an expansion of Halliday's semantics in order to highlight the importance of textual considerations in semantic analysis.

6   'Mounting complexity' is agnate (systemically related) to 'complexity mounts', not to 'x mounts complexity' (as in 'attitude to mounting exhibitions').

7   Halliday does not deploy a stratified model of context in his work, preferring a single metafunctionally organized level of field, tenor and mode analysis when deconstructing context of situation.

8   Halliday in fact uses the term register for the linguistic expression of this level, not for the field, mode and tenor variables themselves — which he refers to collectively as context of situation.

9   Arguments in favour of stratification can be found in Martin, 1992, Chapter 7.

10 Introduced as 'by my friend, Mike Peters, on the social possibilities traditionally open to men and to women.' Gould, 1989:32.

11 Introduced as 'by the cartoonist Szep, on the proper place of terrorism.' Gould, 1989:32.

12 We should note here that Darwin himself never actually wrote these words.

13 In other words, the non-random sequence of mutually expectant graphemes.

14 This clause could be taken as a material (doing) processes, describing scientists' taxonomizing activity. The agent here is however an abstraction, *'folk classifications'*, not a scientist; the processes has accordingly been taken as relational (i.e., possessive/identifying, with an extra agent).

15 With this clause type we are entering another grey area between material (doing) and relational (being) processes; the example is taken as material here because in this sense the process arise takes [present in present] rather than [present] tense in examples such as 'New species are arising during this period . . .'.

16 'Synonymy' has been loosely interpreted here to gather together terms with approximately the same meaning in the text.

17 Projections of this kind provide relatively clear examples of intertextual relations which do not have to be interpreted as mediated by system; as such, within the framework of the system/process theory introduced here, they are the exception rather than the rule.

18 In SFL system and realization is construed probabilistically, in order to build an inflection of semogenesis into the picture, and to blur in one respect the reification of meaning potential inherent in synoptic analysis.

Part 1

Professional Literacy: Construing Nature

Introduction

The first chapter in Part 1, 'On the Language of Physical Science', begins by examining the grammatical features of a typical sentence of modern scientific English. These are then placed in a historical context, going back to the writings of the founders of experimental science. From Newton to the present day there has steadily evolved a form of clause construction characterized, not by *objectivity* as in the popular idea of scientific discourse (which was a late nineteenth-century refinement), but by *objectification* — that is, representing actions and events, and also qualities, as if they were objects. As a corollary to this, the relations *between* events came to be construed as if they were the events themselves: thus instead of 'we did this, then that happened' the scientist writes 'this action of ours was followed by that event'. The reasons why the favourite clause type changed in this way become clear when we take account of the context — both the linguistic context (that of the surrounding discourse) and the social and political context, the way science has evolved in recent centuries in the west.

This kind of grammatical metaphor, whereby a phenomenon of one kind is construed in a way that typically represents a phenomenon of some other kind (actions, events, qualities as if they were things) is probably the major source of problems facing those who are apprenticed to the discourse of science, whether they come to this discourse with English as their mother tongue or with English as a second language. Most of the other linguistic problems they face are related to grammatical metaphor in one way or another: for example complex technical taxonomies, and a high concentration of content words (lexical density). These problems are discussed and illustrated in the second chapter.

Among the writings of major scientific figures in the English-speaking world since Newton's day certain works stand out for the influence they have had on the development of knowledge. One of these is Darwin's *Origin of the Species*. Darwin waited a long time before publishing his findings, which he knew were contentious and would be reviled and misunderstood. In the chapter 5, following a brief historical sketch which picks up the motifs of the third Chapter, Halliday has focused on a key passage in the *Origin of the Species*, the final two paragraphs in which Darwin sums up his ideological stance and reaffirms the rectitude of his own conclusions. A grammatical analysis of the texture of these two paragraphs shows how their effect is achieved.

The first three chapters of this part, taken together, show scientific English at work; in doing so they adopt three complementary perspectives. The first takes typical patterns and treats them historically, as an evolving social semiotic; the second takes problematic examples and treats them developmentally, from the perspective of a learner; the third takes one highly valued passage and treats it as a critical moment in the unfolding of a particular text.

In the sixth chapter language is problematized as being both a part of nature, physical, biological and social, and at the same time a metaphorical construction of the nature of which it is a part. Scientific language has largely reconstrued experience, in ways which scientists themselves now find one-sided and distorting: it creates a massive disjunction between everyday commonsense knowledge and the systematized knowledge of the disciplines. On the other hand, the grammar of everyday language offers an alternative construal of reality which seems more in tune with scientists' current perceptions (as well as being less elitist) — something which becomes especially relevant at a time when communicative (that is, semiotic) systems, of which language serves as prototype, are increasingly invoked as models for understanding physical and biological processes.

The final chapter in Part 1 offers a brief comparison of scientific English and scientific Chinese. Our book as a whole is concerned with English; but it is important to give some indication of how far the features of scientific English may be shared by scientific registers in other languages. It seems likely that nominalizing metaphors in grammar are common to all languages which are used in the construction of science as currently understood in the late twentieth century. Other features differ more or less randomly; thus, Chinese loses more information in the process of nominalizing than English does, but is more explicit in the way it constructs technical taxonomies. Any language, of course, is capable of being evolved as a resource for doing science; the greater the cultural distance, as in any other such semogenic operation, the more work there is to be done.

Other works of Halliday's that are relevant to the chapters in Part 1 include 'New ways of meaning', concerned with the ecological significance of grammar; 'Systemic grammar and the concept of a "science of language"'; and 'Poetry as scientific discourse', on a scientific text of a different kind — lines from Tennyson's 'In Memoriam'. A further sketch of the development of the grammar of scientific English from Chaucer to the present day has been written and will appear in another publication.

Chapter 3

On the Language of Physical Science*

M.A.K. Halliday

The term 'scientific English' is a useful label for a generalized functional variety, or register, of the modern English language. To label it in this way is not to imply that it is either stationary or homogeneous. The term can be taken to denote a semiotic space within which there is a great deal of variability at any one time, as well as continuing diachronic evolution. The diatypic variation can be summarized in terms of field, tenor and mode: in field, extending, transmitting or exploring knowledge in the physical, biological or social sciences; in tenor, addressed to specialists, to learners or to laymen, from within the same group (e.g., specialist to specialist) or across groups (e.g., lecturer to students); and in mode, phonic or graphic channel, most congruent (e.g., formal 'written language' with graphic channel) or less so (e.g., formal with phonic channel), and with variation in rhetorical function — expository, hortatory, polemic, imaginative and so on. So for example in the research programme in the linguistic properties of scientific English carried out at University College London during the 1960s the grid used was one of field by tenor, with three subject areas (biology, chemistry and physics) by three 'brows', high, middle and low (learned journals, college textbooks, and magazines for the general public).

This space-time variation in no way distinguishes scientific English from other registers. A register is a cluster of associated features having a greater-than-random (or rather, greater than predicted by their unconditioned probabilities) tendency to co-occur; and, like a dialect, it can be identified at any delicacy of focus (Hasan, 1973). Whatever the focus, of course, there will always be mixed or borderline cases; but by and large 'scientific English' is a recognizable category, and any speaker of English for whom it falls within the domain of experience knows it when he sees it or hears it.

In this paper I propose to focus on the physical sciences, and to adopt a historical perspective — one which in turn will restrict me to the written mode, since we have had no access to spoken scientific English until very recently. I shall look mainly at material that was written (in its time) for specialists; but seeing the specialist not as a pre-existing persona but as someone brought into being by the discourse itself. I shall concentrate on what seems to me to be the prototypical syndrome of features that characterizes scientific English; and what I hope to suggest is that we can explain how this configuration evolved —

* This chapter is taken from Mohsen Ghadessy (Ed.), *Registers of Written English*, Pinter, 1988, Chapter 10.

provided, first, that we consider the features together rather than each in isolation; and secondly, that we are prepared to interpret them at every level, in lexicogrammatical, semantic, and socio-semiotic (situational and cultural) terms.

• • •

Let us begin with a short example:

> The rate of crack growth depends not only on the chemical environment but also on the magnitude of the applied stress. The development of a complete model for the kinetics of fracture requires an understanding of how stress accelerates the bond–rupture reaction.
> In the absence of stress, silica reacts very slowly with water. (Michalske and Bunker, 1987, p. 81)

Here are instances of some of the features that form part of the syndrome referred to above:

1 The expression **rate of growth**, a nominal group having as Head/Thing the word **rate** which is the name of an attribute of a process, in this case a variable attribute: thus **rate** agnate to **how quickly?**;

2 the expression **crack growth**, a nominal group having as Head/Thing the word **growth** which is the name of a process, agnate to **(it) grows**; and as Classifier the word **crack** which is the name of an attribute resulting from a process, agnate to **cracked** (e.g., **the glass is cracked**), as well as of the process itself, agnate to **(the glass) has cracked**; **crack growth** as a whole agnate to **cracks grow**;

3 the nominal group **the rate of crack growth**, having as Qualifier the prepositional phrase **of crack growth**; this phrase agnate to a qualifying clause **(the rate) at which cracks grow**;

4 the function of **the rate of crack growth** as Theme in the clause: the clause itself being initial, and hence thematic, in the paragraph;

5 the finite verbal group **depends on** expressing the relationship between two things, '*a* depends on *x*': a form of causal relationship comparable to **is determined by**;

6 the expression **the magnitude of the applied stress**: see points 1 and 3 above; its function as culminative in the clause (i.e., in the unmarked position for New information);

7 the iterated rankshift (nominal group in prepositional phrase in nominal group in . . .) in **the development [of [a complete model [for [the kinetics [of [fracture]]]]]]**;

8 the finite verbal group **requires** expressing the relationship between two things, **development . . . requires . . . understanding** (see point 5 above);

9 the parallelism between **(rate of) growth . . . depends on . . . (magnitude of) stress** and **development . . . requires . . . understanding**, but contrasting in that the former expresses an external relationship (third person, *in rebus*: 'if (this) is stressed, (that) will grow'), while the latter expresses an internal relationship (first-and-second person, *in verbis*: 'if (we) want to model, (we) must understand') (see Halliday and Hasan, 1976, p. 240);

10 the expression **an understanding of how** ..., with the noun functioning as Head/Thing being the name of a mental process: agnate to **(we) must understand**; and with the projected clause **how stress accelerates** ... functioning, by rank shift, in the Qualifier;

11 the clause **stress accelerates the bond-rupture reaction**, with finite verbal group **accelerates** as the relationship between two things which are themselves processes: one brings about a change in an attribute of the other, agnate to **makes ... happen more quickly**;

12 the simple structure of each clause (three elements only: nominal group + verbal group + nominal group / prepositional phrase) and the simple structure of each sentence (one clause only);

13 the relation of all these features to what has gone before in the discourse.

To pursue the last point more fully we should have to reproduce a lengthy passage of text; but the following will make it clear what is meant. Prior to **the rate of crack growth** we had had, in the preceding five paragraphs (citing in reverse order, i.e., beginning with the nearest): **speed up the rate at which cracks grow**, **will make slow cracks grow**, **the crack has advanced**, and **as a crack grows**. If we now go right back to the initial section of the text, in the second paragraph we find (**the mechanism by which**) **glass cracks**, (**the stress needed to**) **crack glass**, and (**the question of how**) **glass cracks**; and if we pursue the trail back to the title of the paper, **The Fracturing of Glass**. The title, in other words, is a technical nominalization involving grammatical metaphor; in the text, the metaphor is constructed step by step, (glass) cracks — to crack (glass) — a crack (grows) — the crack (has advanced) — (make) cracks (grow) — (rate of) crack growth. We might predict that later on in the text we would find **crack growth rate**, and indeed we do: **we can decrease the crack growth rate 1,000 times**. Thus the text itself creates its grammar, instantially, as it goes along.

Whenever we interpret a text as 'scientific English', we are responding to clusters of features such as those we have been able to identify in this short paragraph. But it is the combined effect of a number of such related features, and the relations they contract throughout the text as a whole, rather than the obligatory presence of any particular ones, that tells us that what is being constructed is the discourse of science.

• • •

Let me now attempt to give a very brief sketch of how these features evolved. In doing so I shall refer to text examples from various periods; the passages cited are typical of the texts in question, whereas the texts themselves would have been at the frontier of their genre at the time.

In 1391 Chaucer wrote what is now known as his *Treatise on the Astrolabe*, explaining the workings of this instrument to his son Lewis, to whom he had given it as a present on his tenth birthday. In this treatise we find the two first steps towards nominalized discourse: (i) technical nouns, which are either parts of the astrolabe or geometric and mathematical abstractions (such as **latitude, declinacioun, solsticioun**), for example (I.17):

The plate under thy riet ['grid'] is descryved ['inscribed'] with 3 principal cercles; of whiche the leste ['smallest'] is cleped ['called'] the cercle of Cancer, by-cause that the heved ['head'] of Cancer turneth evermor consentrik up-on the same cercle. In this heved of Cancer is the grettest declinacioun northward of the sonne. And ther-for is he cleped the Solsticioun of Somer; whiche declinacioun, aftur Ptholome, is 23 degrees and 50 minutes, as wel in Cancer as in Capricorne.

and (ii) nominal groups with iterated phrase-and-group Qualifier, especially in the more mathematical passages; for example

the latitude [of [any place [in [the region]]]] is the distance [from [the zenith]] [to [the equinoctial]]

The favoured clause types are either relational, as in the earlier passage (attributive for assigning properties, identifying for definitions), or material and mental — these latter in giving instructions, and hence typically imperative as in (II.17):

Tak the altitude of this sterre whan he is on the est side of the lyne meridional, as ney as thou mayst gesse; and tak an assendent a-non right ['straight ahead'] by som maner sterre fix which that thou knowest; and for-get nat the altitude of the firste sterre, ne thyn assendent. And whan that this is don, espye diligently whan this same firste sterre passeth any-thing the south westward, and hath him a-non right in the same noumbre of altitude on the west side of this lyne meridional as he was caught on the est side;

Temporal and causal-conditional clause complexes are formed with 'when', 'if', 'because' (hypotactic), and with 'for', 'therefore' (paratactic); the causal ones are used particularly in explaining why something is called what it is. There is also another kind of hypotactic clause complex, a form of non-defining relative that is rare in modern English; an example was **which declination, after Ptolemy . . .** above (and cf. **the names of the stars are written in the margin of the grid where they are located; of which stars the small point is called the centre**). This is used for tracking an entity from one step in the text to another.

Chaucer's *Treatise* represents a kind of technical, perhaps proto-scientific discourse which is received into English from classical Greek via classical and medieval Latin. It contains technical nouns, both concrete–technological and abstract–scientific; extended nominal groups, especially mathematical; clause complexes which carry forward the argument, of the form '*a*, so/then *x*' or '. . . *b* . . .; which *b* . . .'; and clauses expressing two main fields: the events under study, process type typically relational, for definitions and attributions; and the activity of doing science (using the astrolabe), process types material and mental, for doing and observing + thinking. These are the lexicogrammatical motifs of a text in which scientific English is being conceived.

• • •

For registering the birth of scientific English we shall take Newton's *Treatise on Opticks* (published 1704; written 1675–1687). Newton creates a discourse of experimentation; in place of Chaucer's instructions for use he has descriptions of action — not 'you do this' but 'I did that'. The clauses here are again material, for doing, and mental, for observing and reasoning (**I held/stopped/removed the Prism; I looked through the Prism upon the hole**); and the observations now frequently project, as in **I observed the length of its refracted image to be many times greater than its breadth, and that the most refracted part thereof appeared violet.**

Sample text (Experiment 4):

> In the Sun's Beam which was propagated into the Room through the hole in the Window-shut, at the distance of some Feet from the hole, I held the Prism in such a Posture, that its Axis might be perpendicular to that Beam. Then I looked through the Prism upon the hole, and turning the Prism to and fro about its Axis, to make the Image of the Hole ascend and descend, when between its two contrary Motions it seemed Stationary, I stopp'd the Prism, that the Refractions of both sides of the refracting Angle might be equal to each other, as in the former Experiment. In this situation of the Prism viewing through it the said Hole, I observed the length of its refracted Image to be many times greater than its breadth, and that the most refracted part thereof appeared violet, the least refracted red, the middle parts blue, green and yellow in order. The same thing happen'd when I removed the Prism out of the Sun's Light, and looked through it upon the hole shining by the Light of the Clouds beyond it. And yet if the Refraction were done regularly according to one certain Proportion of the Sines of Incidence and Refraction as is vulgarly supposed, the refracted Image ought to have appeared round.
>
> So then, by these two Experiments it appears, that in Equal Incidences there is a considerable inequality of Refractions. But whence this inequality arises, whether it be that some of the incident Rays are refracted more, and others less, constantly, or by chance, or that one and the same Ray is by Reflection disturbed, shatter'd, dilated, and as it were split and spread into many diverging Rays, as *Grimaldo* supposes, does not yet appear by these Experiments, but will appear by those that follow.

Such descriptions often come in the passive, as in **the Sun's Beam which was propagated into the Room through the hole in the Window-shut, one and the same Ray is by Reflection disturbed, shatter'd, dilated, and as it were split and spread into many diverging Rays.** Note that these have nothing to do with the 'suppressed person' passive favoured by modern teachers and scientific editors, which came into fashion only late in the nineteenth century. They are simply the passive in its typical function in English: that of achieving the balance of information the speaker or writer intends — often describing the result of an experimental step, where the Theme is something other than the Actor in the process (**the Ray . . . is shatter'd**). If the discourse context requires Actor as Theme Newton displays no coyness about using **I**.

When describing the results of an experiment Newton often uses intricate clause complexes involving both expansion and projection, of the form 'I observed that, when I did *a*, *x* happened'. The mathematical sections, on the other hand, display the complementary type of complexity: a single clause with only three elements, but very long and complex nominal groups, as in the final two paragraphs of the following (Experiment 8):

> I found moreover, that when Light goes out of Air through several contiguous refracting Mediums as through Water and Glass, and thence goes out again into Air, whether the refracting Superficies be parallel or inclin'd to one another, that Light as often as by contrary Refractions 'tis so corrected, that it emergeth in Lines parallel to those in which it was incident, continues ever after to be white. But if the emergent Rays be inclined to the incident, the Whiteness of the emerging Light will by degrees in passing on from the Place of Emergence, become tinged in its Edges with Colours. This I try'd by refracting Light with Prisms of Glass placed within a Prismatick Vessel of Water. Now those Colours argue a diverging and separation of the heterogeneous Rays from one another by means of their unequal Refractions, as in what follows will more fully appear. And, on the contrary, the permanent whiteness argues, that in like Incidences of the Rays there is no such separation of the emerging Rays, and by consequence no inequality of their whole Refractions. Whence I seem to gather the two following Theorems.
> 1. The Excesses of the Sines of Refraction of several sorts of Rays above their common Sine of Incidence when the Refractions are made out of divers denser Mediums immediately into one and the same rarer Medium, suppose of Air, are to one another in a given Proportion.
> 2. The Proportion of the Sine of Incidence to the Sine of Refraction of one and the same sort of Rays out of one Medium into another, is composed of the Proportion of the Sine of Incidence to the Sine of Refraction out of the first Medium into any third Medium, and of the Proportion of the Sine of Incidence to the Sine of Refraction out of that third Medium into the second Medium.

Each of the two numbered paragraphs consists of one clause; the verbal groups are **are** and **is composed of**, each having one huge nominal on either side of it. Contrast the first sentence beginning **I found moreover, ...** , where the nominal groups are very simple, but the structure of the sentence, as a clause complex, is highly intricate:

$$\alpha \ \hat{}\ \text{`(}^\times\beta(1\ \hat{}\ +2)\ \hat{}\ \alpha(^\times\beta(1\ \hat{}\ +2)\ \hat{}\ \alpha(\alpha\ \hat{}\ {}^\times\beta(\alpha\ \hat{}\ {}^\times\beta))))$$

What about the technical terms? These fall under five main headings:

general concepts, e.g., Light, Colour, Ray, Beam, Image, Axis
field, specific (optical), e.g., Incidence, Refraction, Medium
field, general (mathematical), e.g., Proportion, Excess, Sine
apparatus and its use, e.g., Prism, Lens, Superficies, Vessel
methodology, e.g., Experiment, Trial, Theorem.

These seem to be a simple extension of what we found in Chaucer. When we look more closely, however, we find something rather different happening. Some of the nouns are words like **emergence**, **whiteness**, **inequality**, **propagation**, which are not within the realm of the technical but are the names of processes or of attributes (agnate to **emerge**, **white**, **unequal**, **propagate**); they are often printed without a capital letter. Let us consider one example:

> Now those Colours argue a diverging and separation of the hetero-geneous Rays from one another by means of their unequal Refractions, . . .

Why does Newton use nouns to refer to processes (**diverging**, **separation**) which are not part of the technical taxonomies? — instead of writing **those Colours argue that the heterogeneous Rays diverge and separate from one another**, or even (since we and not the colours do the arguing) **from those colours we could argue** ('infer') **that . . .**

To explain this grammatical metaphor we have to look at the context, which is the paragraph from **I found moreover . . .** down to **. . . their whole Refractions**. By nominalizing in this way, Newton is achieving two important discoursal effects:

1   packaging a complex phenomenon into a single semiotic entity, by making it one element of clause structure, so that
2   its rhetorical function — its place in the unfolding argument — is rendered fully explicit.

What is this rhetorical function? Or rather, what are these rhetorical functions? — since there are in fact two functions in question, related to each other but distinct. One is the function of Theme, defined in terms of a Theme + Rheme structure; the other is that of New, defined in terms of a Given + New structure.

The Theme is the element that constitutes the point of departure for the message; this is signalled, in English, by first position in the clause. Provided the thematic element is also Given (i.e., non-New), the rhetorical effect is that of *backgrounding*.

The New is the element that constitutes the point of information for the message; this is signalled, in English, by nuclear prominence in the tone group. Provided the informational element is also Rheme (i.e., non-Theme), the rhetorical effect is that of *foregrounding*.

Usually, the pattern of mapping of Theme + Rheme and Given + New on to one another is of this unmarked kind: the Theme is something that is given, and the New is something that is rhematic. This is especially true in written English, where (since there can be no tonic prominence until it is read aloud) the assumption is that the New matter will come in its unmarked position, namely at the end of a clause.

Where the Theme is also Given, and thus typically refers to something that has gone before, it performs a powerful cohesive function in a text: 'you remember what I said just now? — well we're going to move on from there'. This is obviously essential to scientific discourse. But 'what I said just now' is often likely to be the summation of a fairly complex argument, as in the result of Newton's experiment by which he showed that **. . . the Light . . . that . . . emergeth continues ever after to be white**. Newton cannot repeat the whole of this as it stands because it could not form a component part of a new clause; so he packages it into a nominalization **the Whiteness of the emerging Light**, which he can then make thematic. The element is in this way 'backgrounded' as a point of departure.

In the next sentence he also wants to present a rather complex argument, but this time having the complementary status of New. So again he uses this kind of nominalized packaging: **... a diverging and separation of the heterogeneous Rays ... by means of their unequal Refractions**. This is put in culminative position in the clause and hence is interpreted as having tonic prominence. The element is in this way 'foregrounded' as a point of information.

Thus the device of nominalizing, far from being an arbitrary or ritualistic feature, is an essential resource for constructing scientific discourse. We see it emerging in the language of this period, when the foundations of an effective register for codifying, transmitting and extending the 'new learning' are rapidly being laid down.

If so much of the lexical content is nominalized, what is left over for the verb? In **those Colours argue ...**, what Newton is treating as the 'process' is the act of reasoning; or rather, since it is the Colours that are doing the arguing, the relationship of proof that he is setting up between his experimental results and his conclusions. This is one of two motifs that are typically represented by a verb: proving, showing, suggesting and the like. The other motif that is treated in this way is the relationship that is being set up between the processes themselves, e.g., by the verb **arises** in Experiment 4: **... it appears, that in Equal Incidences there is a considerable inequality of Refractions. But whence this inequality arises, ...** It may be easier to discuss this by reference to another example taken from elsewhere in Newton's writings

> The explosion of gunpowder arises therefore from the violent action whereby all the Mixture ... is converted into Fume and Vapour.

The clause again contains two nominalized processes: one backgrounded, **The explosion of gunpowder** (if Newton had written **Gunpowder explodes because ...**, this would have had only **gunpowder**, not its exploding, as Theme); the other foregrounded, **the violent action whereby ...** (enabling the whole of 'the mixture is violently converted into fume and vapour' to be packaged into a single element). But in this instance the verb, **arises (from)**, expresses the relationship between these two processes: one is caused by the other.

In other words, what is being set up as the 'process', by being represented as a verb, is in fact a relation *between* processes: either external '*a* causes *x* to happen', or internal '*b* causes me to think *y*'. Of course, cause is not the only relationship that can be expressed in this way; eventually most of the major categories of expansion come to be represented as verbs (i.e., exposition, exemplification, clarification; addition, variation; time, space, manner, cause, condition, concession; (Halliday, (1985a), pp. 202–16). But cause may have been the one that led the way.

This pattern, of a one-clause sentence consisting of process$_1$ (nominal group) + relation (verbal group) + process$_2$ (nominal group / prepositional phrase), has not yet taken over; it is just coming into prominence. The typical motifs of the *Opticks*, together with their lexicogrammatical realizations, could perhaps be summarized as follows:

1 descriptions of experiments: intricate clause complexes; very little grammatical metaphor; abstract nouns as technical terms of physics;

2 arguments and conclusions from these: less intricate clause complexes; some nominalizations with grammatical metaphor; abstract nouns as non-technical terms (typically processes or attributes);

3 mathematical formulations: clause simplexes ('simple sentences' of one clause only): typically of the form '$a = x$', where a, x are long lexically dense nominal groups with multiple group + phrase embedding; abstract nouns as mathematical technical terms.

But this is already a form of discourse in which the textual organization of the clause, as a movement from a backgrounded 'this is where we are' to a foregrounded 'this is where we are going', has become a powerful resource for the construction and transmission of knowledge.

• • •

There will be space to pause once more along the journey to the present day, to look at the language used by a scientist writing some fifty years after Isaac Newton, namely Joseph Priestley. Priestley's *The History and Present State of Electricity, with Original Experiments* was published in three volumes in the 1760s. This sense of **electricity** meaning 'the study of electricity', is less familiar today, except perhaps in the collocation **electricity and magnetism**; in Priestley's work it is one of an already large number of derivatives of **electric** using the borrowed resources of Graeco-Latin morphology: **electricity, electrical, electrify, electrification, electrician** (a researcher, not someone who comes to mend the wiring); and there is a wealth of terms built up from these, such as **electric light, electric fire** (also not in the modern sense!), **electric fluid, electric circuits, electrical battery, electrical experiment; excited electricity, communicative electricity, medical electricity, conductor of electricity; positive and negative electricity,** (compare **let a person be electrified negatively**), **electric shock** and so on. (Electric shocks were regularly administered in the treatment of paralytic conditions such as tetanus; and also as a pastime, to be transmitted along a human chain — in one instance stretched across the river Thames!) The importance of these terms is that they now begin to form a complex lexical taxonomy, for a defined branch of physics known as 'electricity'.

Meanwhile the grammar has continued to develop along the lines we have already identified. Here is a typical passage, following the section heading **The Theory of Positive and Negative Electricity:**

According to this theory, all the operations of electricity depend upon one fluid *sui generis*, extremely subtle and elastic, dispersed through the pores of all bodies; by which the particles of it are as strongly attracted, as they are repelled by one another.

When the equilibrium of this fluid in any body is not disturbed; that is, when there is in any body neither more nor less of it than its natural share, or than that quantity which it is capable of retaining by its own attraction, it does not discover itself to our senses by any effect. The action of the rubber upon an electric disturbs this equilibrium, occasioning a deficiency of the fluid in one place, and a redundancy of it in another.

The equilibrium being forcibly disturbed, the mutual repulsion of the particles of the fluid is necessarily exerted to restore it.

Let us track two motifs through this piece of discourse. (1) The first paragraph contains a description, which we could modify slightly as follows: the particles of the fluid are as strongly attracted by the pores as they are repelled by one another. In the next paragraph, this is summarized in the form of an abstract technical term **equilibrium**, which functions as Head of a nominal group **the equilibrium of this fluid in any body**, this nominal group functioning as Theme in a clause which is also thematic. Equilibrium is now established as a thing, which can be maintained, disturbed and restored; and the argument can proceed. (2) One component of this equilibrium is that the particles of the fluid are repelled by one another. This too is then picked up and backgrounded: **the mutual repulsion of the particles of the fluid**. In the earlier formulation, the Theme is **the particles of the fluid**; what is news is that they are repelled by one another. This is no longer news, but is now to be taken for granted so as to lead on to some further news, in this case the effect it has in restoring the equilibrium; so it has to be packaged as a single Theme, and this can be achieved only by nominalization, so that **repel + one another** is reworded as **mutual repulsion**, with **the particles of the fluid** as its Qualifier. This is the name of a happening; and the verbal group, **is exerted**, simply tells us that it happens.

By this complex grammatical metaphor, the process of repelling has been reworded to look like an object: repulsion. Under this pressure of the discourse, the nominal elements in the clause are gradually taking over the whole of the semantic content, leaving the verb to express the relationship between these nominalized processes. After this point has been elaborated, the following paragraph uses the same device to present an alternative theory:

Some of the patrons of the hypothesis of positive and negative electricity conceive otherwise of the immediate cause of this repulsion. They say that, as the dense electric fluid, surrounding two bodies negatively electrified, acts equally on all sides of those bodies, it cannot occasion their repulsion. Is not the repulsion, say they, owing rather to an accumulation of the electric fluid on the surfaces of the two bodies; which accumulation is produced by the attraction of the bodies, and the difficulty the fluid finds in entering them? This difficulty in entering is supposed to be owing, chiefly, to the *air* on the surface of bodies, which is probably little condensed there;

Every sentence in that extract would serve to illustrate the point; let us take just the one beginning **Is not the repulsion . . . .** If we 'unpack' this grammatical metaphor we might arrive at some wording such as:

Do not [the electric atmospheres] repel each other because electric fluid has accumulated on the surfaces of the two bodies, [which in turn is] because the bodies are attracted and the fluid cannot easily enter them?

But when the happenings are expressed congruently, as verbs (**repel, accumulated, attract**), the discourse patterning is lost; we no longer have the appropriate thematic and informational movement, the periodicity of background-ing and foregrounding. The metaphorical variant, by using nouns, gives these processes an explicit value with respect to each other in the temporal progression of the discourse; and by a further metaphor uses verbs to construct their semantic interdependency: **occasion, is owing to, is produced by**. The whole con-figuration is an immensely powerful resource for the semiotic construction of reality.

• • •

It is not that these grammatical resources were invented by scientific writers. What the scientists did was to take resources that already existed in English and bring them out of hiding for their own rhetorical purposes: to create a discourse that moves forward by logical and coherent steps, each building on what has gone before. And the initial context for this was the kind of argumentation that was called for by the experimental method in physical science.

Here is a brief summary of the features we have taken into account:

1 Nominal elements:
   — form technical taxonomies
      (a)   technological categories
      (b)   methodological categories
      (c)   theoretical categories
   — summarize and package representations of processes
      (a)   backgrounding (given material as Theme)
      (b)   foregrounding (rhematic material as New)
2 Verbal elements:
   — relate nominalized processes
      (a)   externally (to each other)
      (b)   internally (to our interpretation of them)
   — present nominalized process (as happening)

In other words: concepts are organized into taxonomies, and constructions of concepts (processes) are packaged into information and distributed by back-grounding and foregrounding; and since the grammar does this by nominal-izing, the experiential content goes into nominal groups. The verbal group signals that the process takes place; or, more substantively, sets up the logical relationship of one process to another, either externally (*a* causes *x*), or internally (*b* proves *y*).

By the end of the eighteenth century this has emerged as the most highly-valued model for scientific writing. Two very brief examples; one from John Dalton's *A New System of Chemical Philosophy* (1827):

Hence increase of temperature, at the same time as on one account it increases the absolute quantity of heat in an elastic fluid, diminishes the quantity on another account by an increase of pressure.

one from James Clark Maxwell, *An Elementary Treatise on Electricity* (1881):

The amount of heat which enters or leaves the body is measured by the product of the increase or diminution of entropy into the temperature at which it takes place. [. . .] The consequences which flow from this conjecture may be conveniently described by an extension of the term 'entropy' to electric phenomena.

By this time we find a very large number of different verbs in the functions of (external) **is measured by** and (internal) **may be described by** in this last example. In terms of transitivity, those expressing external relations are relational and either intensive ('be' type, e.g., **be, become, form, equal, represent, constitute, symbolize, signal, herald, reflect, mean, serve as, act as, embody, define, manifest**) or circumstantial ('be' + a circumstantial relation 'at, on, after, with, because of, in order to etc.' e.g., **cause, lead to, accompany, follow, produce, dictate, stimulate, demand, require, correspond to, apply to, arise from, flow from, cover, result from, be associated with, be measured by**). We might include with the latter, in the sense of 'cause' (*a* causes *x*), a number of verbs expressing the causing of a specific effect, e.g., **speed up, encourage, obscure, improve, diminish** ('make faster, more likely, less clear, better, less/fewer'); these can be interpreted, in a grammar of English, either as relational or as material processes, but it is usually the relational feature that predominates when they are used in scientific contexts. There are also the verbs which merely assert that there *is* a process, as in Priestley's **repulsion . . . is . . . exerted**; compare **rapid bonding occurs, considerable momentum develops**.

The verbs expressing internal relations are those such as **prove, show, predict, illustrate, suggest, attest, be explained by, indicate, confirm**. These may also be interpreted as relational intensive, and this interpretation is appropriate when the nominal elements are both abstractions, as in (*Scientific American*, Michalske and Bunker, December 1987, p. 80):

Griffith's energy–balance approach to strength and fracture also suggested the importance of surface chemistry in the mechanical behaviour of brittle materials.

But many of these same verbs also function as sources of projection, as in (ibid., p. 85):

Our discovery of the importance of molecular diffusion near the crack tip indicates that surface coatings might be designed to block the opening of the crack. . . .

Here **indicate** could be interpreted as a mental process 'makes us think that' (compare expressions where the projecting process is itself nominalized, e.g., **leads us to the conclusion that . . .**); while the ones that are most clearly functioning as mental or verbal processes are those where the projection is personalized, e.g., (ibid., p. 80):

Griffith also determined that the smaller the initial crack in a piece of glass is, the greater the applied stress must be to extend it.

But as scientific discourse has come to be depersonalized, during the past hundred years or so, personal projections have tended to be increasingly hedged around: **Smith suggested that** . . . was replaced by **Smith's suggestion was that** . . ., and then by **Smith's suggestion that** . . . followed by some other verb as process (e.g., **is confirmed by**, **conflicts with**); while **I suggested that** . . . has disappeared almost entirely. However, in their more relational functions (including impersonal projections as in **our results show that** . . .) these verbs play a central part in the syndrome of scientific English, constructing the internal steps in the argument whereby a process is paired with one that is evidence for it rather than with one that is its cause.

The thirteen features that we identified in the *Scientific American* text used as illustration at the beginning can all be seen as different manifestations of this underlying pattern which has been developing over the past four to six centuries. During this period, the grammar of scientific English has been continuously evolving; and we have traced this evolution by showing what is the preferred format for representing and explaining physical phenomena. This has changed, through time,

1   externally:
     from *a* happens; so *x* happens
           because *a* happens, *x* happens
              that *a* happens causes *x* to happen
                 happening *a* causes happening *x*
                 to happening *a* is the cause of happening *x*

2   internally:
     from *a* happens; so we know *x* happens
           because *a* happens, we know *x* happens
              that *a* happens proves *x* to happen
                 happening *a* proves happening *x*
                 to happening *a* is the proof of happening *x*

This is, of course, a highly schematic interpretation; but it shows the direction of change. The latest step to date, taken in the twentieth century, is the one whereby the causal (or other) relation itself comes to be nominalized, as in **the cause**, **the proof** above. Let us take a different example and track it backwards up the 'external' arrow of time. The following is from *Scientific American* (Hamilton and Mahrun, July 1986, p. 77):

> The resolution of the experimental difficulties associated with producing and probing exotic atomic nuclei came in the form of an on-line isotope-separation system (or ISOL).

To make it manageable we will leave out the embedded clause (which would have to be tracked as well in its own right). Here is the original together with a possible four regressive rewordings:

> the resolution of the experimental difficulties came in the form of an ISOL
> the experimental difficulties were resolved by the use of an ISOL
> our using an ISOL resolved the difficulties of the experiment

by using an ISOL we solved the difficult parts of the experiment
we used an ISOL and thus could experiment even where it was difficult

Note that these do not vary much in length; despite a common belief, the more nominalized constructions are not, in fact, noticeably shorter.

All these grammatical formats may of course coexist in one paper. Instances of the latest type can be found quite early on — but they are rare. Likewise, a modern scientific article does not remain locked into the most metaphorical wordings from beginning to end; the discourse shifts within the space that grammatical metaphor defines. But there has been a steady drift towards the nominalizing region; and there will be few sentences that do not contain some of the features that we have recognized as its characteristic.

●　●　●

I have tried not just to describe how scientific English has evolved but also to suggest how to explain it. Physical scientists led the way in expanding the grammar of the language, as they found it, so as to construct a new form of knowledge; based on components that were already present in the medieval semiotic — technology on the one hand, and theory on the other — but that had not previously been combined (except perhaps in the long forgotten practice of Roger Bacon). Up to that point, doing and thinking remain as separate moments in the cultural dynamic; in 'science', the two are brought together. This process leaves room for different models of how the two are to be interrelated, which gives rise to currents of thought in humanist philosophy; but it is the practice, the activity of 'doing science', that is enacted in the forms of the language, and there has been a broad consensus about what constitutes scientific practice. It is this reality that is construed in scientific discourse.

Is this form of language more complex? Not necessarily; it depends how we define complexity. If we take lexical density (the number of lexical words per clause), and the structure of the nominal elements (nominal groups and nominalizations), it undoubtedly is more complex. On the other hand, if we consider the intricacy of the sentence structure (the number of clauses in the sentence, and their interdependencies), then it will appear as simpler: mainly one-clause sentences; and likewise with the clause structure — usually only two or three elements in the clause. We are unlikely to find anything as complex as the first sentence of Newton's Experiment 8 in scientific writing today. (Where we will find it is in casual, spontaneous speech.)

It is, however, a language for the expert; one which makes explicit the textual and logical interconnections but leaves many local ambiguities. The ambiguities arise especially in two places: (1) in strings of nouns, leaving inexplicit the semantic relations (mainly transitivity relations) among them; and (2) in the relational verbs, which are often indeterminate and may face both ways (e.g., **higher productivity means more supporting services**: does **means** mean 'brings about', 'is brought about by' or 'requires'?). Here is an example of both, from a Year 6 science textbook:

Lung cancer death rates are clearly associated with increased smoking.

What is **lung cancer death rates**: how quickly lungs die from cancer, how many people die from cancer of the lung, or how quickly people die if they have it? What is **increased smoking**: more people smoke, or people smoke more? What is **are associated with**: caused by (you die because you smoke), or cause (you smoke because you are — perhaps afraid of — dying)? We may have rejected all but the 'right' interpretation without thinking — but only because we know what it is on about already.

Because it is a language for the expert, it can often be problematic for a learner. This is partly a developmental matter: as we have seen, scientific English is highly metaphorical, in the sense of grammatical metaphor, and children find it hard to deal with grammatical metaphor until they reach about secondary-school age. So for children learning science the patterns we have been investigating present a problem in their own right. Apart from this, however, they are faced with a form of language which, while they must use it to construe a whole new realm of experience, tends to leave implicit precisely the experiential meanings that they most depend on for its construction.

To see the language from the point of view of a learner, and especially if one hopes to intervene in the learning process, it is important to understand how it works. For this one needs a 'grammatics' (a model of grammar); I have used systemic grammar, this being the sort of task to which its paradigmatic–functional design is particularly appropriate. (For the framework of analysis, see Halliday, 1985a, 1985c; for a detailed systemic treatment of the grammar of modern scientific English, Huddleston *et al.*, 1968; for scientific English in school, Chapter 8 in the present volume.) But 'how it works' is only part of the story. A newly evolving register is always functional in its context (whether the context itself is one of consensus or of conflict); the language may *become* ritualized, but it cannot start that way, because to become ritualized a feature must first acquire value, and it can acquire value only by being functional. Thus despite the extent to which scientific English comes to be ritualized, and carried over as a language of prestige and power into other contexts where its special features make no sense *except* as ritual (for example in bureaucratic discourse), all the characteristics that we observed, as contributing to the syndrome that was illustrated at the beginning of the paper, are in origin functional in the effective construction of reality, whatever we may feel about the way they are deployed today. And it is this that our 'grammatics' has to be able to account for. Systemic grammar enables us to ask why scientific English evolved the way it did, and how it was able to provide the semiotic base for the emergence of physical science.

Chapter 4

Some Grammatical Problems in Scientific English*

M.A.K. Halliday

In any typical group of science students there will be some who find themselves in difficulty — who find the disciplines of physics, or biology, or mathematics forbidding and obscure. To such students, these subjects appear decidedly unfriendly. When their teacher tries to diagnose the problems the students are having, it is usually not long before the discussion begins to focus on language. Scientific texts are found to be difficult to read; and this is said to be because they are written in 'scientific language', a 'jargon' which has the effect of making the learner feel excluded and alienated from the subject-matter.

This experience is not confined to those who are studying their science in English. It often happens in other languages also that scientific forms are difficult to understand. But here I shall be concentrating on English; and it is important to stress that it is not only ESL, (English as a Second Language) students who find problems with scientific English — so also do many for whom English is the mother tongue. My impression is that, while these two groups — those for whom English is mother tongue and those for whom it is second language — may respond to scientific English in different ways, it is largely the same features that cause difficulties to both. For example, a pile-up of nouns as in **form recognition laterality patterns**, or **glass crack growth rate**, is hard to understand both for ESL and for EL1 (English as a First Language) students of science. The two groups may use different strategies for decoding these structures; but decoding strategies vary according to other factors also, for example the age of the learner. In so far as 'scientific English' presents special problems of its own, distinct from those of other varieties of English, the problems seem to be much the same for everybody.

In any case, in today's multilingual cities such as Birmingham, Toronto or Sydney, there is no clear line between first and second-language groups of learners. A typical secondary-level science class may include monolingual English speakers at one end, students who have had almost no experience of English at the other end, with the remainder spread out all the way along the continuum in

* This chapter is taken from the *Australian Review of Applied Linguistics: Genre and Systemic Functional Studies*, 1989, Series 5, **6**, pp. 13–37.

69

between. In this situation the teacher is forced to think of the problem in terms which apply to all. But this perspective is also relevant to countries such as those of south and south-east Asia, where the students will have been taught using a variety of different languages as their medium of instruction.

Once their attention has been directed on to the language, science teachers usually think of the difficulties first in lexical terms: that is, as difficulties of vocabulary. This is what is implied by the term 'jargon', which means a battery of difficult technical terms. The word 'jargon' often carries a further implication, namely that such terms are unnecessary and the same meaning could have been conveyed without them, in the everyday language of ordinary commonsense. And this is, in fact, one view of scientific language: some people think that it is an unnecessary, more or less ritualistic way of writing, and that science — scientific concepts and scientific reasoning — could just as well be expressed in everyday, non-technical terms. They refer to this other kind of language as 'plain English', 'simple words' and the like.

We could contrast this view with the opposite opinion, which is that science is totally dependent on scientific language: that you cannot separate science from how it is written, or rewrite scientific discourse in any other way. According to this view, 'learning science' is the same thing as learning the language of science. If the language is difficult to understand, this is not some additional factor caused by the words that are chosen, but a difficulty that is inherent in the nature of science itself. It is the subject-matter that is the source of the problem.

Usually when sensible people can hold such opposite points of view, the reality lies somewhere in between; and this is certainly the case in this instance. It would not be possible to represent scientific knowledge entirely in commonsense wordings; technical terms are not simply fancy equivalents for ordinary words, and the conceptual structures and reasoning processes of physics and biology are highly complex and often far removed, by many levels of abstraction, from everyday experience. Hence the language in which they are constructed is bound to be difficult to follow. At the same time, it is often made more difficult than it need be; the forms of scientific discourse can take over, imposing their own martial law, so that writers get locked into patterns of writing that are unnecessarily complicated and express themselves in highly technical wording even in contexts where there is no motive for it. This is the point where we can justifiably talk about 'scientific jargon': where the writer is following a fashion by which he seeks (unconsciously, in all likelihood) to give extra value to his discourse by marking it off as the discourse of an intellectual elite.

It is important to arrive at a balanced view on this question, because we not only need to identify what the problematic features of scientific English are; we also need to try and explain them — to show what functions these things have in the discourse as a whole, and why they have evolved as part of the language of science. This will help us to know whether, in any particular passage, the features that made it difficult to understand were motivated or not — in other words, whether there is some good reason why the text has been written the way it is. Might it be precisely where the complexity is not motivated — where there was no reason for the writer to have adopted that particular wording at that stage in the argument — that the students are finding difficulties? It will take careful, well-informed classroom research to enable us to answer this last question; but we can suggest some explanations, of a general kind, for why these problematic features are found in scientific writing. The language of science, however much it may become a matter of convention, or a way of establishing the writer's own prestige and authority, is not, in origin, an arbitrary code.

But in order to understand why scientific writing became difficult in certain ways, we shall need to get rid of our obsession with words. The difficulty lies more with the grammar than with the vocabulary. In the last resort, of course, we cannot separate these from each other; it is the total effect of the wording — words and structures — that the reader is responding to, and technical terms are part of this overall effect. Nevertheless technical terms are not, in themselves, difficult to master; and students are not particularly dismayed by them. It is usually the teacher who puts technical terms in the centre of the picture, because vocabulary is much more obvious, and easier to talk about, than grammar. But the generalizations we have to make, in order to help students cope with scientific writing, are mainly generalizations about its grammar. The problems with technical terminology usually arise not from the technical terms themselves but from the complex relationships they have with one another. Technical terms cannot be defined in isolation; each one has to be understood as part of a larger framework, and each one is defined by reference to all the others.

I shall suggest seven headings which can be used for illustrating and discussing the difficulties that are characteristic of scientific English:

1 interlocking definitions
2 technical taxonomies
3 special expressions
4 lexical density
5 syntactic ambiguity
6 grammatical metaphor
7 semantic discontinuity

This should not be taken as a definitive listing of categories; all these features could be organized in different ways, or subdivided further, and more could certainly be added. These are simply the headings that I have found useful as a framework for working on the problem. In what follows, I have drawn on various sources, but particularly on the work of my colleagues in Sydney: Charles Taylor's (1979) study of the language of high-school textbooks, with special reference to the problems of second-language learners; Martin and Rothery's (1986) discussion of writing in primary schools; Wignell, Martin and Eggins' (Chapter 8) analysis of geography textbooks at junior and secondary level; and Louise Ravelli's (1985) treatment of grammatical metaphor. My own analysis of scientific texts, reported on in a lecture series at the National University of Singapore, included material from four different points of origin: secondary and upper-primary science and mathematics textbooks from Australia; science lectures recorded at the University of Birmingham in England; writings from the *Scientific American*; and for a historical survey, works by Chaucer, Newton, Priestley, Dalton, Darwin and Clerk Maxwell. I found it necessary to undertake this kind of historical study in order to investigate how, and especially why, the features that were causing such problems of understanding today had themselves originally evolved.[1]

Figure 4.1: Interlocking Definitions of Five Technical Terms

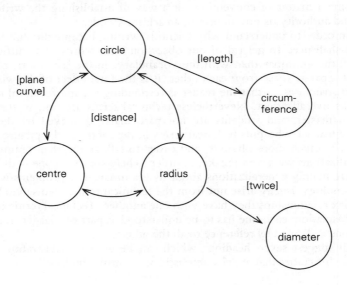

Interlocking Definitions

Here is an example of how a series of definitions is presented to children in upper-primary school:[2]

> A circle is a plane curve with the special property that every point on it is at the same distance from a particular point called the **centre**. This distance is called the **radius** of the circle. The **diameter** of the circle is twice the radius. The length of the circle is called its **circumference**.

Here **circle**, **centre**, **radius**, **diameter** and **circumference** all figure in a series of interlocking definitions. Within this set, **circle**, **centre** and **radius** are mutually defining: they are all used to define each other, through the intermediary of two other terms which are assumed to be already known, namely **distance** and **plane curve**. The remaining terms, **diameter** and **circumference**, are then defined each by reference to one of the first three; and here two other terms are assumed to be known and mastered, namely **length** and **twice**. The pattern of definitions is as in Figure 4.1. Now, there are certain difficulties here which are specific to this example: the notions of 'plane curve', of 'every point on a curve', and of 'the length of a circle'. Likewise, any example chosen would probably present special problems of its own. But at the same time the overall semantic structure is strikingly complex; and this is something that may be found anywhere in maths and science textbooks. The learner has first to reach an understanding of a cluster of related concepts, all at the same time, and then immediately use this understanding in order to derive more concepts from the first ones. Note that these relationships are set up by means of a grammatical construction

Figure 4.2:   Kinds of Climate (Superordination)

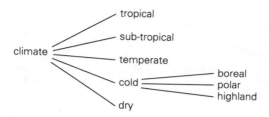

which faces both ways: '*a* is defined as *x*', '*x* is called *a*' — both of which may occur in the same clause, as happens in the first sentence of the extract:

'*a* is defined as an *x* which has feature *y* which is called *b*'

Furthermore the 'hinge' element *y* is itself fairly complex grammatically:

with the special property that every point on it is at the same distance from a particular point

Thus while a technical term poses no great problem in itself — there is nothing difficult about the *word* **diameter**, and its definition **twice the radius** is easy enough to understand *provided you know what the radius is* — a technical *construction* of this kind, in which the terms interlock and are used to define each other, does present the learner with a considerable intellectual task. Writers sometimes try to make the task simpler by adding further definitions, not realizing that in a construct of this kind the greater the number of things defined the harder it becomes to understand.

Technical Taxonomies

These are related to the last heading; but the complexity is of a different kind. In the natural sciences, technical concepts have little value in themselves; they derive their meaning from being organized into taxonomies. Such taxonomies are not simply groups of related terms; they the highly ordered constructions in which every term has a definite functional value. As Wignell, Martin and Eggins point out below in their study of the language of high school geography, a technical taxonomy is typically based on two fundamental semantic relationships: '*a* is a kind of of *x*' (superordination) and '*b* is a part of *y*' (composition). Thus in their example of **climate**, climate is divided into certain *kinds* (Figure 4.2), and is composed of certain *parts* (Figure 4.3). It will be seen that the first is an 'either/or' relationship: 'every climate is *either* tropical *or* subtropical *or* . . .'; the second is a 'both + and' relationship: 'every climate is *both* temperature *and* solar radiation *and* . . .'. (We have to stretch the meaning of **either** and **both** here so that they are no longer limited to just two.)

Figure 4.3: *Parts of Climate (Composition)*

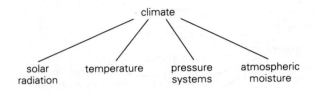

Three problems can arise with such constructions. The first is that these taxonomies can become very complicated, with many layers of organization built into them. The second is that they are usually not made explicit; there are often neither lists nor diagrams (the figures above do not appear in the textbook), so the student is left to work them out for himself from reading the text. The third problem is that the criteria on which these taxonomies are set up can also be extremely complex, so that they need to be described and explained in some detail.

It would be possible to make the reading matter more learner-friendly by dealing systematically with these three problems in turn: first introducing the terms in their taxonomic order (e.g., **there are five kinds of climate, namely . . .**), then setting them out in lists or diagrams, and finally describing each category and, where possible, explaining it. In practice, the first and third steps are usually taken together, with the second one being left out; as a result, the way the taxonomy is presented is often grammatically very confusing, with no clear pattern of theme and information running through it. For example,

ONE-CELLED ORGANISMS. Some organisms, such as the ameba and others in the culture you examined, are composed of only one cell. These organisms are said to be **unicellular**. Living in water, these animals are in close contact with the food, water, and oxygen they need. A one-celled animal takes in its own food. Along with this food, the animal also takes in some water. Additional water enters the animal cell by diffusion. The normal movement of the cytoplasm carries the food, water, and oxygen throughout the cell. Waste materials are eliminated directly to the outside of the cell. Most one-celled organisms can survive only in a watery environment.

It is very likely that the writer of this passage has been trying to make it more interesting for the reader by varying the order and the manner of presenting the categories to be learnt: the kinds of organism, the parts of the organism and so on. Thus every clause begins with a new theme: **some organisms, these organisms, living in water, a one-celled animal, additional water, the normal movement of the cytoplasm, waste materials, most one-celled organisms**. Unfortunately, while this kind of variation may be an admirable goal for a literary text, if scientific texts are written in this way they are much harder to read and to learn from. It is very difficult to construct the relevant taxonomies on the basis of this kind of writing.

Special Expressions

Some expressions used in mathematical language have a special grammar of their own, for example **solving the open sentence over D**. Here it is the expression as a whole that gets to be defined, rather than any particular word in it:

> If *D* is the domain of a variable in an open sentence, the process of finding the truth set is called **solving the open sentence over D**.

This is 'technical grammar', rather than technical terminology; it is not particularly problematic once it has been explained (provided the learner does not ask what happens if *D* is *not* the domain of a variable in an open sentence).

This kind of special grammar is more common in mathematics than in science; mathematicians have often had to stretch the grammar a little in order to say what they want. Already in Isaac Newton's writings we find some very long nominal constructions, like the following from the *Treatise on Optics*:

> The Excesses of the Sines of Refraction of several sorts of Rays above their common Sine of Incidence when the Refractions are made out of divers denser Mediums immediately into one and the same rarer Medium, suppose of Air, . . .

— all of which is merely the Subject of the clause. This kind of stretching of the grammar is less usual in scientific discourse. However, the language of science has brought its own innovations, stretching the grammar in ways which are at first sight less obvious but which, partly because they are less obvious, tend to cause greater difficulties of comprehension. Here is an example from an upper primary-school textbook:

> Your completed table should tell you what happens to the risk of getting lung cancer as smoking increases.

The **table** is, of course, a table of figures; that is understood. But how does a table **tell you** something? — tables do not talk, even tables of figures. And what kind of an object is a **risk**, such that we can ask **what happens** to it? And what does **smoking increases** mean: that more smoke is put out by some combustion process? What kind of relationship is being expressed by the **as**: does it mean 'while' (time), 'because' (cause), or 'in the same way that' (manner)?

What is being illustrated here is not, in face, a single phenomenon. It is a set of interrelated phenomena: features which tend to go together in modern scientific writing, forming a kind of syndrome by which we recognize that something is written in the language of science. But although these features commonly go together, in order to understand the problems they pose to a student we will need to separate them out; and this will occupy the next three headings. The present section will serve as a bridge leading into them, because when we see them in their historical perspective they do constitute a special mode of expression that evolved in scientific discourse, although we are now so used to them that we no longer think of them as special. It is only when they occur in a fairly extreme form that they stand out, as in the following (taken from an abstract):[3]

[These results] are consistent with the selective perceptual orientation hypothesis if it is assumed that both word recognition and concurrent verbal memory produce more left than right hemisphere activation and that in the case of mixed lists in the present study this activation had not dissipated on form recognition trials.

Lexical Density

This is a measure of the density of information in any passage of text, according to how tightly the lexical items (content words) have been packed into the grammatical structure. It can be measured, in English, as the number of lexical words per clause.

In the following examples, each of which is one clause, the lexical words are in bold type; the lexical density count is given at the right:

(a) But we never did anything very much in **science** at our **school**. 2
(b) My **father** used to **tell** me about a **singer** in his **village**. 4
(c) A **parallelogram** is a **four-sided figure** with its **opposite sides parallel**. 6
(d) The **atomic nucleus absorbs** and **emits energy** in **quanta**, or **discrete units**. 8

In any piece of discourse there is obviously a great deal of variation in the lexical density from one clause to the next. But there are also some general tendencies. In informal spoken language the lexical density tends to be low: about two lexical words per clause is quite typical. When the language is more planned and more formal, the lexical density is higher; and since writing is usually more planned than speech, written language tends to be somewhat denser than spoken language, often having around four to six lexical words per clause. But in scientific writing the lexical density may go considerably higher. Here are three clauses with a lexical density of 10–13, all from *Scientific American* (December, 1987):

(e) Griffith's **energy balance approach** to **strength** and **fracture** also **suggested** the **importance** of **surface chemistry** in the **mechanical behaviour** of **brittle materials**. 13
(f) The **conical space rendering** of **cosmic strings'** **gravitational properties applies** only to **straight strings**. 10
(g) The **model rests** on the **localized gravitational attraction exerted** by **rapidly oscillating** and extremely **massive closed loops** of **cosmic string**. 13

When the lexical density goes up to this extent, the passage becomes difficult to read. Of course, the difficulty will also depend on the particular lexical items that are used and on how they are distributed in the grammatical structure; but the lexical density is a problematic factor in itself. In much scientific writing, almost all the lexical items in any clause occur inside just one or two nominal groups (noun phrases); compare examples (e)–(g) above, where this applies to all except one in each case (**suggested, applies, rests**). Perhaps the hardest examples to process are those which consist of strings of lexical words without any grammatical words in between, such as **Griffith's energy balance approach, cosmic strings' gravitational properties**; likewise those cited at the beginning of the paper, **form recognition laterality patterns** and **glass crack growth rate**. Even where the words themselves are perfectly simple and well known, as in the last of these four examples, the expressions are not easy to understand. Another example was **the increasing lung cancer death rate**, which appeared in the same passage as the example quoted in the last section. Here, however, another factor contributes to the difficulty, that of grammatical ambiguity; and this leads us in to our next heading.

Syntactic Ambiguity

Consider examples such as the following:

(h)   Increased responsiveness may be reflected in feeding behaviour.
(j)   Lung cancer death rates are clearly associated with increased smoking.
(k)   Higher productivity means more supporting services.

All have a very simple structure: a nominal group, functioning as Subject, followed by a verbal group, followed by another nominal group with (in two instances) a preposition introducing it. If we focus attention on the verbal expressions, **may be reflected (in)**, **are . . . associated (with)**, **means**, we find that they are ambiguous; and they are ambiguous in two respects. In the first place, we cannot tell whether they indicate a relationship of cause or of evidence. Is one thing being said to be the *effect of* another, or is it merely the *outward sign of* it? For example: in (h), does the feeding behaviour *demonstrate that* responsiveness has increased, or does it *change as a result of* the increase? In the second place, supposing that we can identify a relationship of cause, we still cannot tell which causes which. In (k), for example, is higher productivity *brought about by* more supporting services, or does it *cause* more supporting services to be provided? It may seem obvious to the writer, and also to a teacher, which meaning is intended; but it is far from obvious to a learner, and teacher and learner may interpret the passage differently without either of them being aware that another interpretation was possible.

The expression **are associated with**, in (j), can also face in either direction: either 'cause' or 'are caused by'. *We* may know that smoking causes cancer, and hence that the more you smoke, the more likely you are to die from cancer of the lung. But this sentence *could* mean that lung cancer death rates *lead to* increased smoking: perhaps people are so upset by fear of lung cancer that they need to smoke more in order to calm their nerves. It is even possible that the writer wanted not to commit himself to a choice between these two interpretations of the statistics. But when we start to explore the meaning of this example more carefully, we find that it contains a great deal more ambiguity in addition to that which we have already seen in the verb.

For example, what does **lung cancer death rates** mean? Is it 'how many people die from lung cancer', or 'how quickly people die when they get lung cancer'? Or is it perhaps 'how quickly people's lungs die from cancer'? And does **increased smoking** mean 'people smoke more', or 'more people smoke' — or is it a combination of the two, 'more people smoke more'? Having reached some understanding up to this point, such as 'more people smoke . . . more people die of cancer', we still do not know whether they are the same people or not — is it just the smokers who die more, or everyone else as well? Nor do we know whether the situation is real or hypothetical: is it 'because more people are smoking, so more are dying', or 'if more people smoked, more would die'? if we combine all these possibilities we have already reached some fifty possible interpretations, most of which were quite plausible; they are genuine alternatives faced by a human reader, not fanciful simulations of some computerized parsing program.

Where does this ambiguity come from? It arises from various sources. We have already referred to polysemous verbs like **mean, be associated with**; there are probably between 1,000 and 2,000 verbs of this class in use in scientific English. But the main cause of ambiguity is that clauses are turned into nouns. That is to say, something that would in spoken English be typically expressed as a clause is expressed instead as a group of words centring on a noun. If I say **Mary announced that she had accepted**, I am making it clear who did what; but if I say **the announcement of Mary's acceptance**, you cannot tell: whether Mary made the announcement herself or someone else did; whether Mary was accepting (something) or being accepted; whether she had accepted/been accepted already or would accept/be accepted in the future. Thus the single nominal group **the announcement of Mary's acceptance** corresponds to many different wordings in the form of a clause: **Mary announced that she would accept, they announced that Mary had been accepted**, and so on. A great deal of semantic information is lost when clausal expressions are replaced by nominal ones.

Scientific writing uses very many nominal constructions of this kind, typically in combination with verbs of the type illustrated in (h)–(k) above. Both these features are, as we have seen, highly ambiguous, although we usually do not recognize the ambiguity until we try to re-word the passage in some other form. Here is a further example:

(l) The growth of attachment between infant and mother signals the first step in the development of the child's capacity to discriminate amongst people.

Possible rewordings of this might be:

{When / If} an infant and {its / a} mother {start to grow / grow more} attached to one another, {this shows that / this is because} the child {is taking / has taken} the first steps towards {becoming / becoming more} capable of {distinguishing / preferring} one person from/to another.

Combining these we get $2^7 = 128$ possible interpretations. But in this instance I find it difficult to opt for any one of them; none of the rewordings seems to be particularly convincing.

Grammatical Metaphor

The high lexical density and the ambiguity discussed in the last two sections are both by-products of a process I shall refer to as 'grammatical metaphor'. This is like metaphor in the usual sense except that, instead of being a substitution of one *word* for another, as when we say **you're talking tripe** instead of **you're talking nonsense**, it is a substitution of one grammatical class, or one grammatical structure, by another; for example, **his departure** instead of **he departed**. Here the words (lexical items) are the same; what has changed is their place in the grammar. Instead of pronoun **he** + verb **departed**, functioning as Actor plus Process in a clause, we have determiner **his** + noun **departure**, functioning as Deictic plus Thing in a nominal group.[4] Other examples are **her recent speech concerned poverty** instead of **she spoke recently concerning poverty; glass crack growth rate** instead of **how quickly cracks in glass grow**. Often the words may change as well as the grammar, as in the last example where **how quickly** is replaced by **rate** — we do not usually say **glass crack growth quickness**; but the underlying metaphor is in the grammar, and the lexical changes follow more or less automatically.

I am not suggesting that there will always be some absolute, non-metaphorical form to which these grammatical metaphors can be related; metaphor is a natural historical process in language and modes of expression involving different degrees of metaphor will always exist side by side. We can often take two or three or even more steps in rewording a grammatical metaphor in a less metaphorical, more congruent form; for example, we might say that 'cracking' is really a process — something happening — rather than a thing, so that **cracks in glass**, with **cracks** as a noun, is a metaphor for **glass cracks** with **cracks** as verb. As another example,

(m) [The 36 class only appeared on this train] in times of reduced loading, or engine failure.

could be reworded as **when loadings were reduced, or the engine failed**; but we might then reword the first part over again as **when the load was smaller** or ever **when fewer goods were being carried**.

What is the nature of this rewording? One way of thinking of it is by imagining the age of the reader, or listener. In talking to a 9-year-old, we would never say **in times of engine failure**; we would say **whenever the engine failed**. Notice that we have not had to simplify the vocabulary; there are no difficult words in the first version — it is the grammar that is difficult for a child. Similarly we would change **slow down the glass crack growth rate** to **make the cracks in glass grow more slowly**, or **stop the cracks in glass from growing so quickly**. What we are doing, when we reword in this way, is changing the grammar (with some consequential changes in vocabulary) by making it *younger*. Children learn first to talk in clauses; it is only later — and only when they can already read and write with facility — that they are able to replace these clauses with nominal groups.

As far as we can tell, this also reflects what happened in the history of the language. In English, and other languages of Europe, the older pattern is the clausal one; and it is based on certain principles of wording which we might summarize as follows:

1 processes (actions, events, mental processes, relations) are expressed by verbs;
2 participants (people, animals, concrete and abstract objects that take part in processes) are expressed by nouns;
3 circumstances (time, place, manner, cause, condition) are expressed by adverbs and by prepositional phrases;
4 relations between one process and another are expressed by conjunctions.

For example:

parti-cipant	process	circum-stance	relation between processes	participant	process	circumstance
the cast	acted	brilliantly	so	the audience	applauded	for a long time
[noun]	[verb]	[adverb]	[conjunc-tion]	[noun]	[verb]	[prepositional phrase]

If this is now reworded metaphorically as:

the cast's brilliant acting	drew	lengthy applause	from the audience
[noun]	[verb]	[noun]	[prepositional phrase]

a number of changes have taken place. The processes **acted** and **applauded** have been turned into nouns, **acting** and **applause**; the participant **the cast** has become a possessive, while **the audience** has become part of a prepositional phrase. The circumstances **brilliantly** and **for a long time** have both become adjectives inside nominal groups; and the relation between the two processes, showing that one of them caused the other, has become a verb, **drew**. This makes it sound as though acting and clapping were things, and as if the only event that took place was the cause relation between them (. . . **acting drew** . . . **applause**). All these changes illustrate what is meant by grammatical metaphor.

This kind of metaphor is found particularly in scientific discourse, and may have evolved first of all in that context. It is already beginning to appear in the writings of the ancient Greek scientists; from them it is carried over into classical Latin and then into medieval Latin; and it has continued to develop — but to a far greater extent — in Italian, English, French, German, Russian and the other languages of Europe from the Renaissance onwards. And although it has spread across many different registers, or functional varieties, of language, in English at least the main impetus for it seems to have continued to come from the languages of science.

Why did scientific writers, from Isaac Newton onwards, increasingly favour such a mode of expression? — one in which, instead of writing 'this happened, so that happened', they write 'this event caused that event'? These were not arbitrary or random changes. The reason lies in the nature of scientific discourse. Newton and his successors were creating a new variety of English for a new kind of knowledge; a kind of knowledge in which experiments were carried out; general principles derived by reasoning from these experiments, with the aid of mathematics; and these principles in turn tested by further experiments. The discourse had to proceed step by step, with a constant movement from 'this is what we have established so far' to 'this is what follows from it next'; and each of these two parts, both the 'taken for granted' part and the new information, had to be presented in a way that would make its status in the argument clear. The most effective way to do this, in English grammar, is to construct the whole step as a single clause, with the two parts turned into nouns, one at the beginning and one at the end, and a verb in between saying *how* the second follows from the first.

I have written about the history of this development elsewhere, with illustrations from some of the earlier texts (Chapter 3). What I am presenting here is a very simplified account; there are, obviously, countless variations on the pattern described above. Nevertheless these variants all derive from the basic principle of organizing information into a coherent form that suited the kind of argumentation that came to be accepted as 'scientific'. Here is a contemporary example, taken from the *Scientific American*:

> The atomic nucleus absorbs and emits energy only in quanta, or discrete units. Each absorption marks its transition to a state of higher energy, and each emission marks its transition to a state of lower energy.

Notice how, in the second sentence, each clause consists of (i) a 'taken for granted' part, nominalizing what has been said before (**the atomic nucleus absorbs energy → each absorption; the atomic nucleus emits energy → each emission**); (ii) a 'new information' part, pointing forward to what is to come, and also nominalized (**its transition to a state of higher / lower energy**); and (iii) the relation between them, in the form of a verb (**marks**). Frequently the 'taken for granted' part summarizes the whole of a long previous discussion; for example, the same article contains the sentence:

> The theoretical program of devising models of atomic nuclei has of course been complemented by experimental investigations.

This has exactly the same pattern; but here the 'taken for granted' part (**the theoretical program ... atomic nuclei**) is referring back to many paragraphs of preceding text.

If we reword these so as to take the metaphor out, the entire balance of the information is lost. For the last example we might write:

> We devised models of atomic nuclei, in a program of theoretical [research], and in addition of course we investigated [the matter] by doing experiments.

But this would give us to indication that the first part was a summary of what had gone before, or that the last part was going to be taken up and developed in what followed. What is equally important, it would fail to make it clear that each step — devising theoretical models and investigating experimentally — is to be understood as a unity, a single phenomenon rather than an assembly of component parts.

It would be wrong to give the impression that in developing this favourite type of clause structure, and the grammatical metaphor that made it possible, the scientists were guided by any conscious planning. They were not. Newton and his contemporaries did discuss the best ways of constructing a scientific paper, and they tried to regulate the use of vocabulary for building elaborate taxonomies, especially in biology (and taken up later on in chemistry); but they were not aware of their own use of grammar, and these forms evolved naturally in response to pressure from the discourse.[5] It is only when we analyse this discourse grammatically, using a functional grammar, that we can appreciate how the patterns relate to what the scientists were trying to achieve.

I have not presented the detailed grammatical analysis here; it would need too much space. But it is helpful, I think, to bring out the nature of grammatical metaphor, and the sense in which these forms can be said to be metaphorical, because almost every sentence in scientific writing will contain some example of it, and it does present problems to the learner. This is partly a question of maturity: students well into secondary school may still find it difficult to comprehend, even if they have been educated throughout in English medium.[6] For those who are taking up English just as a language for science and technology, the problem may be greater or less depending on the degree and kind of grammatical metaphor found in the language(s) they have used as medium of education before.

It seems likely that part of the difficulty arises, however, because these metaphorical expressions are not just another way of saying the same thing. In a certain sense, they present a different view of the world. As we grew up, using our language to learn with and to think with, we have come to expect (unconsciously, until our teachers started to give us lessons in grammar) that nouns were for people and things, verbs for actions and events. Now we find that almost everything has been turned into a noun. We have to reconstruct our mental image of the world so that it becomes a world made out of things, rather than the world of happening — events with things taking part in them — that we were accustomed to. Some of the problem may even be ideological: the student may want to resist this view of reality that he feels is being imposed on him by the language of science. It is worth noting, in the connections, that the scientists themselves are now becoming dissatisfied with the language they use in their writings. They too feel that it has gone too far in this direction, and that if they are to continue to develop new ideas in science they will need to return to less nominalized forms of expression.[7]

Semantic Discontinuity

This is my final heading; I am using it to point out that writers sometimes make semantic leaps, across which the reader is expected to follow them in order to reach a required conclusion. Let me discuss just one example:

In the years since 1850, more and more factories were built in northern England. The soot from the factory smokestacks gradually blackened the light-coloured stones and tree trunks.

Scientists continued to study the pepper moth during this time. They noticed the dark-coloured moth was becoming more common. By 1950, the dark moths were much more common than the light-coloured ones.

However, strong anti-pollution laws over the last twenty years have resulted in cleaner factories, cleaner countryside and an increase in the number of light-coloured pepper moths.

The first two paragraphs are rather straightforward; but in the third paragraph, problems arise. Taken as a whole, it is a typical example of the structure described in the last section: two processes, with a logical connection between them. The sense is '*a* happened, so *x* happened', expressed metaphorically in the form of 'happening *a* caused happening *x*' (**strong anti-pollution laws . . . have resulted in cleaner factories . . .**). We might reword this part as:

Over the last twenty years, [the government have passed] strong laws to stop [people] polluting; so the factories [have become] cleaner . . .

We saw above that the main reason for choosing the metaphorical form was that 'happening *a*' was something that had been presented before, and so here was being referred to as a whole, as a kind of package or summary of what was to be taken for granted and used as a point of departure for the next step in the argument. However, in this instance happening *a* has not been presented before; this is the first time we have heard of any 'anti-pollution laws'. So the reader has to: discover that it is new information; decode it; and use it as a stepping-off point for understanding something else.

But let us suppose that the reader has coped with this difficult assignment. He now comes to 'happening *x*' and finds that this is a coordination of three processes, all of them presented metaphorically: **cleaner factories, cleaner countryside and an increase in the number of light-coloured pepper moths**. Rewording this, he begins to understand:

. . . the factories have become cleaner, the countryside has become cleaner, and there are more light-coloured pepper moths than before.

— that is, the moths have also become cleaner: only a few of them are now affected by dirt in the air. But that is not at all the intended message. What the reader is supposed to do is to insert another logical relationship between each pair of these resulting processes, and then draw a highly complex conclusion from them:

. . . the factories have become cleaner, [so] the countryside has become cleaner, and [so] there are getting to be more of the light coloured pepper moths [because they don't show up against clean trees, and therefore do not get eaten by the birds as much as they did when the trees were dirty].

In other words, the learner is expected to work out for himself the principle of natural selection.

This is a particularly problematic example. The language is highly metaphorical, in the sense of grammatical metaphor; the first part of the sentence is misleading because it suggests that we know about the 'strong anti-pollution laws' already, and in the second part the reader is required to perfom two complicated semantic leaps — inserting the two causal connectives, and working out the implications of the second one. But it is not uncommon to find semantic discontinuities of one kind or another in scientific writing; the specialist has no trouble with them — but for learners they are an additional hazard. Of all the kinds of difficulty discussed in these few pages, this is the one a teacher can do least towards helping students to solve. The teacher can give a few illustrations, and warn the students to be on their guard; but every instance seems to be unique, and it is hard to find any general principles behind them all.

Conclusion

Most of the features described under these seven headings could in principle occur independently of each other. But they are all closely related, and, excepting perhaps those mentioned under 'Special Expressions' (in mathematics), they tend to cluster together as characteristics of scientific discourse. I have tried to show that they are not arbitrary — that they evolved to meet the needs of scientific method, and of scientific argument and theory. They suit the expert; and by the same token they cause difficulty to the novice. In that respect, learning science is the same thing as learning the language of science. Students have to master these difficulties; but in doing so they are also mastering scientific concepts and principles.

At the same time, it must be said that many of those who write in the language of science write it very badly. They leave implicit things that need to be made explicit, create multiple ambiguities that cannot readily be resolved, and use grammatical metaphor both inappropriately and to excess. The language thus becomes a form of ritual, a way of claiming status and turning science into the prerogative of an elite. Learners who complain that their science texts are unnecessarily difficult to read may sometimes be entirely justified. And we are all familiar with those who, not being scientists, have borrowed the trappings of scientific language and are using it purely as a language of prestige and power — the bureaucracies and technocracies of governments and multinational corporations.[8] In bureaucratic discourse these features have no reason to be there at all, because there is no complex conceptual structure or thread of logical argument. But they serve to create distance between writer and reader, to depersonalize the discourse and give it a spurious air of being rational and objective.

In my view the best tool we have for facing up to this kind of language, criticizing it where necessary but above all helping students to understand it, is a functional model of grammar. This enables us to analyse any passage and relate it to its context in the discourse, and also to the general background of the text: who it is written for, what is its angle on the subject matter, and so on. Grammatical analysis is a fairly technical exercise, and not something that students can be expected to undertake for themselves unless they are specializing in language.

But science teachers (provided they can be persuaded to discard traditional prejudices about grammar!) may find it interesting and rewarding to explore the language of their own disciplines; and also, where this applies, to compare scientific English with scientific registers that have evolved, or are now evolving, in the major languages of the region in which they work.

Notes

1   See entries in the References for Taylor (1979), Martin and Rothery 1986, Ravelli (1985). Primary texts for the historical survey, as indicated in Chapter 3 above, were Geoffrey Chaucer, *A Treatise on the Astrolabe* 1391; Isaac Newton, *Optics* (1704): Joseph Priestley, *The History and Present State of Electricity* (1767); John Dalton, *A New System of Chemical Philosophy* (1827); Charles Darwin, *The Origin of the Species* (1859); James Clerk Maxwell, *An Elementary Treatise on Electricity* (1881). Texts from the *Scientific American* were Hamilton and Maruhn, 'Exotic atomic nuclei' (July 1986); Michalske and Bunker, 'The fracturing of glass', Vilenkin, 'Cosmic strings' (December 1987). For the University of Birmingham studies see King (in prep.). A sketch of some features of the grammar of scientific English is contained in Chapter 3 above; the work from which the present paper is mainly derived was presented in lecture form in Halliday (1986).

2   Sources for the upper-primary / lower-secondary science and mathematics texts quoted in this paper are McMullen and Williams (1971); Intermediate Science Curriculum Study, (1976); Parkes, Couchman and Jones (1978); Vickery, Lake, McKenna and Ryan (1978). The taxonomies of climate are from Sale, Friedman and Wilson (1980).

3   From the Abstract to Hellige (1978).

4   For the analysis of the grammar see Halliday (1985a), Chapters 5 and 6.

5   For the evolution of the scientific article see Bazerman (1988). For an account of the work of the scientific language planners at the time of Newton, see Salmon (1966, 1979).

6   See Lemke (1982, 1983) for the results of a detailed investigation of the teaching of science in American high schools. For discussion of science education in Britain, with reference to the language of science, see White and Welford (1987).

7   This point is discussed briefly in Chapter 6 below.

8   For an analysis of the nature and function of technocratic discourse see Lemke (1990b).

Chapter 5

The Construction of Knowledge and Value in the Grammar of Scientific Discourse: Charles Darwin's *The Origin of the Species**

M.A.K. Halliday

Our text for this symposium is 'verbal and iconic representations: aesthetic and functional values'. I shall start from verbal representations and functional values; but I shall suggest that functional values may be also aesthetic, and verbal representations may be also iconic. The first part of the paper will be a general discussion; then in the second part I shall focus on one particular text, the final two paragraphs of Darwin's *The Origin of the Species*.

I shall assume the concept of *register*, or functional (diatypic) variation in language. It is convenient to talk of 'a register', in the same way that one talks of 'a dialect': in reality, of course, dialectal variation is typically continuous, along many dimensions (that is, with many features varying simultaneously), and what we call 'a dialect' is a syndrome of variants that tend to co-occur. Those feature combinations that actually do occur — what we recognize as 'the dialects of English', for example, or 'the dialects of Italian' — are only a tiny fraction of the combinations that would be theoretically possible within the given language. Similarly, 'a register' is a syndrome, or cluster of associated variants; and again only a small fraction of the theoretically possible combinations will actually be found to occur.[1]

What is the essential difference between dialectal variation and diatypic or register variation?[2] Prototypically, dialects differ in expression; our notion of them is that they are 'different ways of saying the same thing'. Of course, this is not without exception; dialectal variation arises from either geographical conditions (distance and physical barriers) or social-historical conditions (political, e.g., national boundaries; or hierarchical, e.g., class, caste, age, generation and sex), and as Hasan (1990) has shown dialects that are primarily social in origin can and do differ also semantically. This is in fact what makes it possible for dialect variation to play such an important part in creating and maintaining (and also in transforming) these hierarchical structures. Nevertheless dialectal variation is primarily variation in expression: in phonology, and in the morphological formations of the grammar.

* This chapter is taken from Clotilde de Stasio et al. (Eds), *La rappresentazione verbale e iconica*, Milano: Guerini, 1990.

Registers, on the other hand, are not different ways of saying the same thing; they are ways of saying different things. Prototypically, therefore, they differ in content. The features that go together in a register go together for semantic reasons; they are meanings that typically co-occur. For this reason, we can translate different registers into a foreign language. We cannot translate different dialects; we can only mimic dialect variation.

Like dialects, registers are treated as realities by the members of the culture. We recognize 'British English', 'American English', 'Australian English', 'Yorkshire dialect', 'Cockney' etc.; and likewise 'journalese', 'fairy tales', 'business English', 'scientific English' and so on. These are best thought of as spaces within which the speakers and writers are moving; spaces that may be defined with varying depth of focus (the dialect of a particular village versus the dialect of an entire region or nation; the register of high-school physics textbooks versus the register of natural science), and whose boundaries are in any case permeable, hence constantly changing and evolving. A register persists through time because it achieves a contingent equilibrium, being held together by tension among different forces whose conflicting demands have to be met.[3] To give a brief example, grossly oversimplified but also highly typical: what we call 'scientific English' has to reconcile the need to create new knowledge with the need to restrict access to that knowledge (that is, make access to it conditional on participating in the power structures and value systems within which it is located and defined).

In a recent short paper on the language of physical science I set out to identify, describe and explain a typical syndrome of grammatical features in the register of scientific English (see Chapter 3). I cited a short paragraph from the *Scientific American* and focused particularly on the pattern represented in the following two clauses:

The rate of crack growth depends . . . on the chemical environment.
The development of a . . . model . . . requires an understanding of how stress accelerates the bond rupture reaction.

In their most general form, these clauses represent the two related motifs of '*a* causes / is caused by *x*', '*b* proves / is proved by *y*'. Let me cite another pair of examples taken from a different text:

These results cannot be handled by purely structural models of laterality effects . . .   [*b* + prove + *y*]

(if . . .) both word recognition and concurrent verbal memory produce more left than right hemisphere activation.   [*a* + cause + *x*]

Taken together: '*b* cannot be explained by *y* if *a* causes *x*'. At the level of the syntagm (sequence of classes), each of these consists of two nominal groups linked by a verbal group whose lexical verb is of the 'relational' class, in this case **handle, produce**. Their analysis in systemic-functional grammar, taking account of just those features that are relevant to the present discussion, is as set out in Figure 5.1[4].

In that paper I tried to show how and why this pattern evolved to become the dominant grammatical motif in modern scientific English. Historically the

Figure 5.1: Transitivity (Ideational), Mood (Interpersonal), and Theme and Information (Textual) Structures in the 'Favourite' Clause Type

	these results	cannot	be handled	by	purely structural models of laterality effects
transitivity	Value / Identified	Process: relational / circumstantial: cause (internal)			Token / Identifier

mood	Mood		Residue	
	Subject	Finite	Predicator	Adjunct

theme	Theme	Rheme

information	Given	←– – – – – – – – – – – – – – – – . New

	both word recognition and concurrent verbal memory	produce	more left than right hemisphere activation
transitivity	Token / Identified	Process: relational / circumstantial: cause (external)	Value / Identifier

mood	Mood		Residue	
	Subject	Finite	Predicator	Complement

theme	Theme	Rheme

information	Given	←– – – – – – – – – – – – – – – – New

process is one of dialectic engagement between the nominal group and the clause. It is a continuous process, moving across the boundary between different languages: it began in ancient Greek, was continued in classical and then in medieval Latin, and then transmitted to Italian, English and the other languages of modern Europe. Table 5.1 is a summary of the relevant grammatical features that led up to this dominant motif, as they appear in two influential early scientific texts: Chaucer's *Treatise on the Astrolabe* (c. 1390) and Newton's *Opticks* from three hundred years later. What is not found in Chaucer's text, but is found in Newton, is this particular syndrome of clausal and nominal features: a clause of the type analysed in Figure 5.1 above, in which the nominal elements functioning as Token and Value are nominalizations of processes or properties; for example,

Table 5.1:   Some Grammatical Features in the Scientific Writings of Chaucer and Newton

	grammatical features	typical contexts
Chaucer: *Treatise on the Astrolabe*		
1: nominal	nouns:	technical terms:
	noun roots	technological (parts of instrument)
	nouns derived from verbs and adjectives	astronomical and mathematical
	nominal groups (with prepositional phrase and clause Qualifiers)	mathematical expressions
2: clausal	material and mental; imperative	instructions ('do this', 'observe/reckon that')
	relational ('be', 'be called'); indicative	observations; names and their explanations
Newton: *Opticks*		
1: nominal	nouns:	technical terms:
	noun roots	general concepts; experimental apparatus
	nouns derived from verbs and adjectives	physical and mathematical
	nominal groups (with prepositional phrase and clause Qualifiers)	mathematical expressions
	*nominalizations of processes and properties	logical argumentation; explanations and conclusions
2: clausal	material and mental; indicative	description of experiments ('I did this', 'I saw/reasoned that')
	*relational ('cause', 'prove'); indicative	logical argumentation; explanations and conclusions

* = not found in Chaucer's text

> The unusual Refraction is therefore perform'd by an original property of the Rays. (Newton, 1952, p. 358)

This is still very much a minority type in Newton's writing; but it is available when the context demands. In order to see when context does demand it, let me cite the immediately preceding text:

> . . . there is an original Difference in the Rays of Light, by means of which some Rays are . . . constantly refracted after the usual manner, and others constantly after the unusual manner. For if the difference be not original, but arises from new Modifications impress'd on the Rays at their first Refraction, it would be alter'd by new Modifications in the three following Refractions; whereas it suffers no alteration, but is

constant, . . . The unusual Refraction is therefore perform'd by an original property of the Rays.

Note in particular the sequence **[are] constantly [refracted] after the unusual manner . . . The unusual Refraction is therefore perform'd by . . . . .** Formulaically: '*a* happens . . . The happening of *a* is caused by . . .'. The nominalization **the unusual refraction** refers back to the earlier formulation **are refracted after the unusual manner**, in such a way as to make it the startingpoint for a new piece of information explaining how it is brought about.

This grammatical pattern exploits the universal *metafunctional* principle of clause structure: that the clause, in every language, is a mapping of three distinct kinds of meaning — interpersonal, ideational and textual (clause as action, clause as reflection, clause as information). The structural mechanism for this mapping, as it is worked out in English, was shown in Figure 5.1. What concerns us here first and foremost is the textual component. In English the clause is organized textually into two simultaneous message lines, one of Theme + Rheme, and one of Given + New. The former presents the information from the speaker's angle: the Theme is 'what I am starting out from'. The latter presents the information from the listener's angle — still, of course, as constructed for him by the speaker: the New is 'what you are to attend to'. The two prominent functions, Theme and New, are realized in quite distinct ways: the Theme segmentally, by first position in the clause; the New prosodically, by greatest pitch movement in the tone group. Because of the different ways in which the two are constituted, it is possible for both to be mapped on to the same element. But the typical pattern is for the two to contrast, with tension set up between them, so that the clause enacts a dynamic progression from one to the other: from a speaker-Theme, which is also 'given' (intelligence already shared by the listener), to a listener-New, which is also 'rhematic' (a move away from the speaker's startingpoint). This pattern obviously provides a powerful resource for constructing and developing an argument.[5]

We could refer to this in *Gestalt* terminology as a move from 'ground' to 'figure', but that sets up too great a discontinuity between them and I shall prefer the 'backgrounding–foregrounding' form of the metaphor since it suggests something more relative and continuous. The type of clause that is beginning to emerge in the Newtonian discourse, then, constructs a movement from a *backgrounded* element which summarizes what has gone before to a *foregrounded* element which moves on to a new plane. But there has to be a third component of the pattern, namely the relationship that is set up between the two; and it is this that provides the key to the potentiality of the whole, enabling the clause to function effectively in constructing knowledge and value. We have said that the relationship is typically one of cause or proof, as in the examples so far considered (**depends on, accelerates, produce, arises from, is performed by; requires an understanding of, cannot be handled by**). That was an oversimplification, and we now need to consider this relationship a little more closely.

The grammar of natural languages constructs a set of logical-semantic relations: relations such as 'i.e.,', 'e.g.,', 'and', 'or', 'but', 'then', 'thus', 'so'. These are grammaticalized in various ways, typically (in English) by conjunctions and prepositions. There are many possible ways of categorizing these relations, depending on the criteria adopted; one schema that I find useful in applying the model of the grammar to discourse analysis is that shown in Table 5.2.[6]

Table 5.2: Common Types of Logical-Semantic Relation, with Typical Realizations as Conjunction and Preposition

expansion type	category	typical conjunction	typical preposition
1 elaborating	expository exemplificatory	in other words; i.e., for example; e.g.,	namely such as
2 extending	additive alternative adversative	and or but	besides instead of despite (in contrast)
3 enhancing	temporal causal conditional concessive comparative	then (at that time) so (for that reason) then (in that case) yet so (in that way)	after because of in the event of despite (contrary to expectation) like

Table 5.3: Examples of Lexicalization of Logical-Semantic Relations (as Verbs)

category	examples of lexicalization (verbs)
expository exemplificatory	be represent constitute comprise signal reflect be exemplify illustrate
additive alternative adversative	accompany complement combine with replace alternate with supplant contrast with distinguish
temporal causal conditional concessive comparative	follow precede anticipate co-occur with cause produce arise from lead to result in prove correlate with be associated with apply to contradict conflict with preclude resemble compare with approximate to simulate

Note: Verbs in the same category are not, of course, synonymous, since they embody other features such as negative, causative. No distinction is shown here between 'external' (*in rebus*) and 'internal' (*in verbis*).

In the type of clause that we are considering here, however, these relationships come to be lexicalized, as verbs; for example, the verbs **produce, arise from, depend on, lead to** as expressions of the causal relationship. Furthermore, this logical-semantic space is then crosscut along another dimension, according to whether the relationship is being set up *in rebus* or *in verbis*;[7] thus the causal relationship may be either (*in rebus*) 'a causes x' or (*in verbis*) 'b proves (= causes one to say) y'. Not all the logical-semantic relationships are lexicalized to the same extent; nor is this last distinction between relations in the events and relations in the discourse equally applicable to all. But the general pattern is as shown, with the experiential content entirely located within the two nominal groups and the verbal group setting up the relation between them. Table 5.3 lists some of the common verbs by which these logical-semantic relations are construed in lexical form.

Only a handful of these verbs occur in Newton's writings. The number has noticeably increased half a century later, in Joseph Priestley's *History and Present State of Electricity*; and by the time of James Clerk Maxwell's *An Elementary Treatise on Electricity*, after another 100 years, there are some hundreds of them

Table 5.4: Examples Showing Logical-Semantic Relations Lexicalized (1) as Verb, (2) and (3) as Noun

1 Theme	nominal group	the theoretical program of devising models of atomic nuclei
Relation (extending: additive)	verbal group	has of course been complemented by
New	nominal group (noun: process)	experimental investigations
2 Theme	nominal group	the resulting energy level diagram
Relation (elaborating: expositive)	*be* + nominal group (noun: relation)	is in essence a representation of '*b* represents *y*'
New	nominalization (rankshifted clause)	what nature allows the nucleus to do
3 Theme	nominal group (noun: process)	the *resolution* of the experimental difficulties ⊃
Relation (enhancing: causal)	*come* + nominal group (noun: relation)	came in the form of '*a* was resolved by *x*'
New	nominal group	an on-line isotope separation system

Source: J.H. Hamilton and J.A. Maruhn, 'Exotic atomic nuclei', *Scientific American*, July 1986.

in current use. My guess is that in modern scientific writing there are somewhere around 2,000, although in the early twentieth century a countertendency arose whereby the logical-semantic relationship is relexicalized, this time as a noun, and the verb is simply **be** or other lexical lightweight such as **have, bring, need.** The pattern is then '*a* is the cause of *x*', '*b* is the proof of *y*'; thus **is the cause of, is the result of, is a concomitant of, has as a consequence, is a representation of, is an alternative to; is the proof of, needs explanation as, is an illustration of, serves as evidence for**, etc. Table 5.4 displays some examples from a text in the *Scientific American*.

We can appreciate, I think, how such verbal representations are themselves also iconic. (1) There is a movement from a given Theme (background) to a rhematic New (foreground); this movement in time construes iconically the flow of information. (2) New semiotic entities are created by these nominal packages, like **rate of crack growth, left/right hemisphere activation, unusual refraction, resolution of the experimental difficulties**; the nominal expression in the grammar construes iconically an objectified entity in the real world. (3) The combination of (1) and (2) construes iconically the total reality in which we now live, a reality consisting of semiotic entities in a periodic flow of information — a flow that one might well say has now become a flood. The grammar constructs this world, as it has constructed (and continues to construct) other worlds; and it does so, in this case, by this complex of semogenic strategies: 'packaging' into extended nominal groups, nominalizing processes and properties, lexicalizing logical-semantic relations first as verbs and then as nouns, and constructing the whole into the sort of clause we meet with everywhere — not just in academic writing but in the newspapers, in the bureaucracy, and in our school textbooks — typified by the following from a primary-school science text: **lung cancer death rates are clearly associated with increased smoking**. The grammar of a natural language is a theory of experience, a metalanguage of daily life; and the forms of verbal representation that evolved as part of modern science have penetrated into almost every domain of our semiotic practice.

• • •

Let me now move to the second part of the paper, which I realize will appear somewhat detached from the first, although I hope the overall direction will soon become clear. I am still taking as my 'text' the language of science, but now contextualizing it within a more literary frame of reference. I said earlier that the concept of register, as functional variation in language, implies that our domain of inquiry is a text type rather than an individual text; we are interested in what is typical of this or that variety. In stylistics, on the other hand, we have traditionally been interested in the highly valued text as something that is unique, with the aim of showing precisely that it is not like any other texts. There are of course more or less codified genres of literature, text types showing similar text structures such as narrative fiction or lyric poetry; but there is no such thing as a literary register, or 'literary English' as a functional variety of English.

Does this mean that we cannot have a highly valued text in some definable register such as the language of science? Clearly it does not. For one thing, we can treat any text as a unique semiotic object / event. If we take a piece of scientific writing and 'read' it as a work of literature, we locate it in two value systems which intersect in a series of complementarities: (1) between the text as representing a register or type and the text as something unique; (2) between the traditional 'two cultures', scientific and humanistic, the one privileging ideational meaning the other privileging interpersonal; (3) within the scientific, an analogous opposition between (in terms of eighteenth century thought) the uniformity of the system and the diversity of natural processes, or (reinterpreted in modern terms) between order and chaos.[8]

But there are some texts which by their own birthright lie at the intersection of science and verbal art: which are not merely reconstituted in this dual mode by us as readers, but are themselves constituted out of the impact between scientific and poetic forces of meaning. I have written elsewhere about the crucial stanzas of Tennyson's 'In Memoriam,' those which seem to me to lie at the epicentre of one such semiotic impact (Halliday, 1987b). That is a text that would be categorized, in traditional terms, as elegiac poetry but containing certain passages with a scientific flavour or motif; and by studying its grammar we can get a sense of what that implies. In the text I am concerned with here, this relationship of 'science' to 'literature' is reversed: *The Origin of the Species* will be classified in the library under 'science', whereas in certain lights it appears as a highly poetic text. Interestingly, while in the Tennyson poem this impact is most strongly felt at a point more or less halfway through the text, here it is most striking at the very end, — in the final two paragraphs, according to my own reading of the book. Text 1 reproduces the two paragraphs in question.

Here Darwin not only sums up the position for which he has been arguing (over some 450 pages, in my 1979 edition) but also defends it against the

Text 1
from Charles Darwin, *On the Origin of the Species by means of Natural Selection* (1859)
London: J. Murray,

Authors of the highest eminence seem to be fully satisfied with the view that each species has been independently created. To my mind it accords better with what we know of the laws impressed on matter by the Creator, that the production and extinction of the past and present inhabitants of the world should have been due to secondary causes, like those determining the birth and death of the individual. When I view all beings not as special creations, but as the lineal descendants of some few beings which lived long before the first bed of the Silurian system was deposited, they seem to me to become ennobled. Judging from the past, we may safely infer that not one living species will transmit its unaltered likeness to a distant futurity. And of the species now living very few will transmit progeny of any kind to a far distant futurity; for the manner in which all organic beings are grouped, shows that the greater number of species of each genus, and all the species of many genera, have left no descendants, but have become utterly extinct. We can so far take a prophetic glance into futurity as to foretell that it will be the common and widely-spread species, belonging to the larger and dominant groups, which will ultimately prevail and procreate new and dominant species. As all the living forms of life are the lineal descendants of those which lived long before the Silurian epoch, we may feel certain that the ordinary succession by generation has never once been broken, and that no cataclysm has desolated the whole world. Hence we may look with some confidence to a secure future of equally inappreciable length. And as natural selection works solely by and for the good of each being, all corporeal and mental endowments will tend to progress towards perfection.

It is interesting to contemplate an entangled bank, clothed with many plants of many kinds, with birds singing on the bushes, with various insects flitting about, and with worms crawling through the damp earth, and to reflect that these elaborately constructed forms, so different from each other, and dependent on each other in so complex a manner, have all been produced by laws acting around us. These laws, taken in the largest sense, being Growth with Reproduction; Inheritance which is almost implied by reproduction; Variability from the indirect and direct action of the external conditions of life, and from use and disuse; a Ratio of Increase so high as to lead to a Struggle for Life, and as a consequence to Natural Selection, entailing Divergence of Character and the Extinction of less-improved forms. Thus, from the war of nature, from famine and death, the most exalted object which we are capable of conceiving, namely, the production of the higher animals, directly follows. There is grandeur in this view of life, with its several powers, having been originally breathed into a few forms or into one; and that, whilst this planet has gone cycling on according to the fixed law of gravity, from so simple a beginning endless forms most beautiful and most wonderful have been, and are being, evolved.

opposition and ridicule which he knew it was bound to evoke. The initial clause in the last sentence of all, **There is grandeur in this view of life,** presents a defiant, if perhaps rather forlorn, challenge to those whose only after-image of the text would be (as he foresaw) the humiliation of finding that they were descended from the apes.

I shall offer a very partial grammatical analysis of these paragraphs, taking account just of the two features discussed above: the 'textual' organization of the clause in terms (1) of Theme – Rheme and (2) of Given – New. In embarking on this analysis, I was interested in finding out what rhetorical or discursive strategies Darwin was using, as he summed up his case and worked up to the resounding climax of that final clause: **from so simple a beginning endless**

forms most beautiful and most wonderful have been, and are being, evolved. The patterning is not at all obvious; to me, at least, it did not stand out on the surface of the text. On the contrary, perhaps; one thing that makes this passage so effective may be that the reader is not presented with any explicit signal that 'this is the nature of my argument'.

Why then did I think that the clause by clause analysis of Theme and of New would be likely to reveal anything of interest? In general, these features of the clause grammar play a significant part in constructing the flow of the discourse. We have seen above, first, how they give texture to a single clause and, secondly, how they construe a pair of clauses into a coherent logical sequence, interacting with referential and lexical cohesion. In addition to this, the ongoing selection of elements functioning as Theme, and elements functioning as New, throughout a portion of a text is a major source of continuity and discursive power. In a seminal article written some ten years ago, on the status of Theme in discourse, Peter Fries (1981/1983) showed that it was possible to relate Theme in the clause to the concepts of 'method of development' and 'main point' in composition theory.[9] Any motif that figured regularly as clause Theme could be seen to function as 'method of development' in the text, while any motif that figured regularly as Rheme was likely to be functioning as 'main point'. Fries was concerned specifically with the category of Theme and so based his interpretation on the straightforward division of each clause into two parts, the Theme and the Rheme, treating the Rheme as equal in prominence with the Theme. This has the advantage with a written text that one does not need to give it the 'implication of utterance', as is necessary if one wants to identify the element that is New. But the category of New is more appropriate, since it identifies prominence that is of a different kind and would therefore be expected to have a distinct function in the discourse; it is also more constraining, since not everything that is outside the Theme will fall within the New.[10] Here therefore I shall take it that what constitutes the 'main point' of the discourse is any motif that figures regularly as New. The third reason for analysing this aspect of the grammar of the text, then, is that the analysis reveals a great deal about the organization of the discourse. All these considerations would of course apply to any text. But in many texts these patterns are near the surface, and emerge very quickly once one begins to read them carefully; whereas here they come to light only when one consciously attends to the grammatical structure.

Text 2 shows the Theme in each ranking clause throughout the text.[11] The grouping of these into motifs is set out in diagrammatic form in Figure 5.2. The first motif that emerges (numbered I in Figure 5.2) is very clearly that of authority, beginning with the Theme of the first clause **authors of the highest eminence**. This, when followed by **seem to be fully satisfied**, becomes solidary with a passage in the final paragraph of an earlier chapter where, mentioning a number of authorities who have (*contra* Darwin) **maintained the immutability of species,** he then goes on: **But I have reason to believe that one great authority, Sir Charles Lyell, from further reflexion entertains grave doubts on this subject**.[12] The motif of authority is thus already given, constructed out of the morphological relationship of **authors = authorities.** Darwin now extends it in a sequence of clause Themes as follows:

authors of the highest eminence — the Creator [to my mind] — I — we

Text 2
the text showing Theme — Rheme structure (Theme italicized)

Authors of the highest eminence seem to be fully satisfied with the view *that each species* has been independently created. *To my mind it* accords better with what we know of the laws impressed on matter by the Creator, that the production and extinction of the past and present inhabitants of the world should have been due to secondary causes, like those determining the birth and death of the individual. *When I* view all beings not as special creations, but as the lineal descendants of some few beings which lived long before the first bed of the Silurian system was deposited, *they* seem to me to become ennobled. *Judging from the past, we* may safely infer *that not one living species* will transmit its unaltered likeness to a distant futurity. *And of the species now living* very few will transmit progeny of any kind to a far distant futurity; *for the manner in which all organic beings are grouped*, shows *that the greater number of species of each genus, and all the species of many genera*, have left no descendants, but have become utterly extinct. *We* can so far take a prophetic glance into futurity as to foretell that it will be *the common and widely-spread species, belonging to the larger and dominant groups*, which will ultimately prevail and procreate new and dominant species. *As all the living forms of life* are the lineal descendants of those which lived long before the Silurian epoch, *we* may feel certain *that the ordinary succession by generation* has never once been broken, and *that no cataclysm* has desolated the whole world. *Hence we* may look with some confidence to a secure future of equally inappreciable length. *And as natural selection* works solely by and for the good of each being, *all corporeal and mental endowments* will tend to progress towards perfection.

It is interesting to contemplate an entangled bank, clothed with many plants of many kinds, with *birds* singing on the bushes, with *various insects* flitting about, and with *worms* crawling through the damp earth, and to reflect *that these elaborately constructed forms*, so different from each other, and dependent on each other in so complex a manner, have all been produced by laws acting around us. *These laws*, taken in the largest sense, being Growth with Reproduction; Inheritance which is almost implied by reproduction; Variability from the indirect and direct action of the external conditions of life, and from use and disuse; a Ratio of Increase so high as to lead to a Struggle for Life, and as a consequence to Natural Selection, entailing Divergence of Character and the Extinction of less-improved forms. *Thus, from the war of nature, from famine and death*, the most exalted object which we are capable of conceiving, namely, the production of the higher animals, directly follows. *There is grandeur* in this view *of life*, with its several powers, having been originally breathed into a few forms or into one; *and that, whilst this planet* has gone cycling on according to the fixed law of gravity, *from so simple a beginning* endless forms most beautiful and most wonderful have been, and are being, evolved.

By this thematic progression Darwin establishes his own claim to authority, wherewith to dispute and override these authors of the highest eminence. He first appeals to the Creator — but being careful to precede this with the interpersonal Theme **to my mind**, which both protects him against the arrogance of claiming to know the Creator's purposes, and, by a neat metafunctional slip (from interpersonal 'me' to ideational 'me'),[13] leads naturally from his role as interpreter of the Creator's design to his position as an authority in his own right. This **I** is then modulated to **we**, in **we can so far take a prophetic glance into futurity**; preceded by a hypotactic clause (also thematic) **judging from the past**, which — without saying who or what is doing the judging (since it is nonfinite and so needs no Subject) — justifies the assumption that 'I' am in fact speaking on behalf of us all. Thus the clause Themes have by this point securely

Figure 5.2: *Motifs Constructed as Theme*

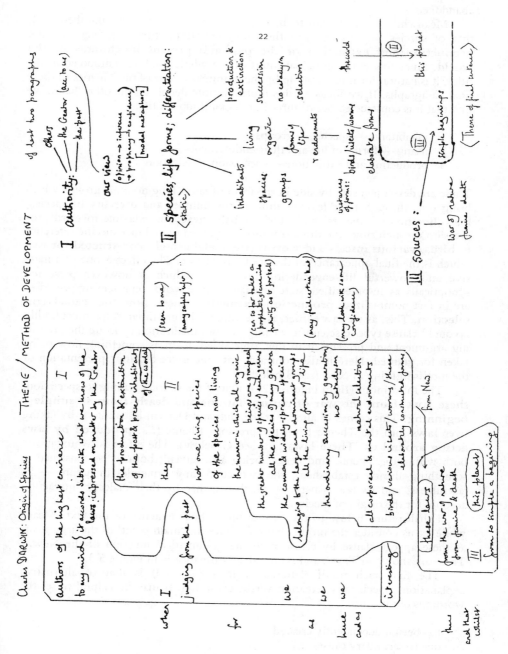

underpinned Darwin's own status as an authority; and this thematic motif is now abandoned.

Meanwhile, it has begun to be overtaken by another motif (numbered II), that of species, life forms, and their differentiation; 'first introduced as **the production and extinction of the past and present inhabitants of the world**. Since this is the principal motif of the whole book (as embodied in the title), it is natural for it to be set up by the grammar as one of the Themes of these final paragraphs. If we focus on this motif in more detail, on the other hand, we find that it is constructed out of three interlocking sub-motifs:

1 inhabitants — species — groups
2 living — forms of life and their endowments
3 production and extinction — succession — selection

These are developed side by side in the form of fairly long nominal groups which bring out, through their lexicogrammar, the number and diversity of species, the collocation of 'species' with 'life', and the steady, irreversible forward progression through time; the three sub-motifs are then united in a specific reference to **birds, various insects** and **worms (these elaborately constructed forms)**, which is the final appearance of this motif as Theme. The effect is one of a massive and powerful life-engendering process — which is however presented synoptically as an objectified 'state of affairs', since words representing processes are in fact nominalized: **production, extinction, succession, (no) cataclysm, selection**. This, as we saw earlier, is a feature of the grammar of the most highly favoured clause type in scientific writing: the nominalization picks up the preceding argument and presents it in this 'objectified' form as something now to be taken for granted. Here it also contrasts with the more dynamic presentation of the motifs figuring as New (see B below).

The third motif (III in Figure 5.2) is that of the sources leading to speciation: **these laws; from the war of nature, famine and death; from so simple a beginning**. This comes in almost at the end; and Darwin leads into it by taking over **laws** into the Theme from the previous Rheme (. . . **produced by laws acting around us. These laws** . . .; see C below). The effect is to juxtapose, both within the Theme (and hence, being also 'given', both to be construed as something already established), the two conflicting principles in nature — its lawfulness, and its lawlessness — which together by their dialectic interaction account for the origin of species.

I shall return below to the extraordinary final sentence of the text. Meanwhile let us consider the motifs that constitute the 'main point' of the argument, as these appear clause by clause with the grammatical function of the New.[14] These are shown in Text 3 and set out diagrammatically in Figure 5.3.

The first such motif (lettered A in Figure 5.3) is that of alternative explanations: specifically, creation versus evolution. It may be helpful to set the wordings out in a list:

has been independently created
due to secondary causes
those [causes] determining the birth and death of the individual
not as special creation (but as . . .)

Text 3
the text showing Given — New structure (New italicized)

Authors of the highest eminence seem to be fully satisfied with the view that each species *has been independently created*. To my mind it accords better with *what we know of the laws impressed on matter by the Creator*, that the production and extinction of the past and present inhabitants of the world should have been *due to secondary causes*, like *those determining the birth and death of the individual*. When I view all beings *not as special creations*, but as *the lineal descendants* of some few beings which lived *long before the first bed of the Silurian system was deposited*, they seem to me to become *ennobled*. Judging from *the past*, we may safely infer that not one living species will transmit *its unaltered likeness to a distant futurity*. And of the species now living very few will transmit *progeny of any kind* to a far distant futurity; for the manner in which all organic beings are grouped, shows that the greater number of species of each genus, and all the species of many genera, have left *no descendants*, but have become *utterly extinct*. We can so far take *a prophetic glance into futurity* as to foretell that it will be the common and widely-spread species, belonging to the larger and dominant groups, which will ultimately *prevail* and procreate *new and dominant species*. As all the living forms of life are *the lineal descendants of those which lived long before the Silurian epoch*, we may feel certain that the ordinary succession by generation *has never once been broken*, and that no cataclysm has desolated *the whole world*. Hence we may look with some confidence *to a secure future of equally inappreciable length*. And as natural selection works solely *by and for the good of each being*, all corporeal and mental endowments will tend to progress *towards perfection*.

It is interesting to contemplate *an entangled bank*, clothed with *many plants of many kinds*, with birds singing *on the bushes*, with various insects *flitting about*, and with worms crawling *through the damp earth*, and to reflect that these elaborately constructed forms, so different from each other, and dependent on each other in so complex a manner, have all been produced *by laws acting around us*. These laws, taken in the largest sense, being Growth with Reproduction; Inheritance which is almost implied by reproduction; Variability from the indirect and direct action of the external conditions of life, and from use and disuse; a Ratio of Increase so high as to lead to a Struggle for Life, and as a consequence to Natural Selection, entailing Divergence of Character and the Extinction of less-improved forms. Thus, from the war of nature, from famine and death, *the most exalted object which we are capable of conceiving*, namely, *the production of the higher animals*, directly *follows*. There is *grandeur* in this view of life, with its several powers, having been originally breathed *into a few forms or into one*; and that, whilst this planet has gone cycling on *according to the fixed law of gravity*, from so simple a beginning *endless forms most beautiful and most wonderful have been, and are being, evolved*.

> the lineal descendants
> long before the first bed of the Silurian system was deposited

The final one of these is the last appearance of this motif until the very last words of the text (. . . **have been, and are being, evolved**); meanwhile, via the two semantic features of generation (**lineal descendants**) and antiquity (**long before the first bed** . . .), it leads us in to the second of the 'New' motifs, that of transmission — or better, transmitting, since the way the grammar constructs it is at least as much clausal as nominal.[15]

Like the second of the thematic motifs, this second motif within the New (B in Figure 5.3) is also constructed out of three sub-motifs:

Figure 5.3: *Motifs constructed as New*

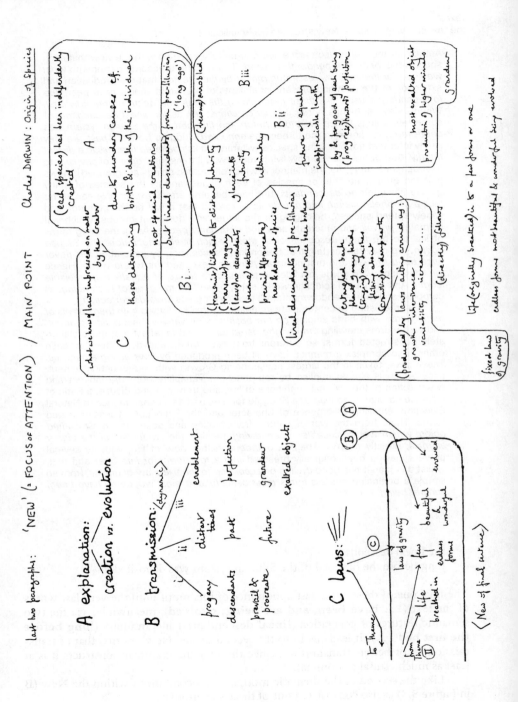

Table 5.5: Summary of Motifs Constructing Theme and New of Ranking Clauses

Theme ('method of development')		*New* ('main point')
I authorities	A	explanation
II the phenomenon of species: 1. inhabitants and groups 2. life forms and endowments 3. variation	B	the process of evolution: 1. transmission and procreation 2. time: past — future 3. ennoblement
III sources (laws/war of nature)	C	law (natural selection)

Note: (in final two paragraphs of *The Origin of the Species*)

1   progeny: (leave) descendants — (become) extinct — (procreate) new and dominant species
2   time: remote past — distant futurity (**a secure future of inappreciable length**)
3   ennoblement: (become) ennobled — (progress) towards perfection — most exalted object — the production of higher animals

The first two of these co-occur; the third is introduced at the beginning of this motif (**become ennobled**), then left aside and taken up again after the sub-motifs of progeny and time have been established. The message line is that descendancy across the ages equals ennoblement, and that this process will continue in the future as it has done in the past. The effect of associating the 'evolutionary' motifs of progeny and time with this one of ennoblement is to collocate evolution with positively-loaded interpersonal expressions like **by and for the good of each being, towards perfection** and so on; this might serve to make such an unpalatable concept slightly less threatening and more acceptable.

There is then a short, transitional motif comprised of the environment in which the diversity of species (the birds, insects etc. of II above) can be appreciated: **an entangled bank, plants of many kinds, (singing) on bushes, flitting about, (crawling) through the damp earth.** This could perhaps be seen as an appendage to B above, illustrating the progress towards perfection; but it is also transitional, via the search for explanation (**have all been produced by**), to the final motif (lettered C in Figure 5.3) which is broached as **laws acting around us.** These laws are then enumerated, as a long list of nominal groups, all with embedded phrases and/or clauses in them and all functioning as the final element in the one ranking clauses — a clause which is (anomalously) non-finite, despite being the main and only clause in the sentence.[16]

Up to this point, then (that is, up to the final sentence of the final paragraph), the clauses are rather clearly organized, through their textual functions of Theme (in Theme — Rheme) and New (in Given — New), around a small number of distinct but interlocking motifs. We could summarize this pattern as in Table 5.5 above. Then, in the final sentence, the motifs of II, III, B and C are all brought together and in an extremely complex pattern. The sentence begins with **There is grandeur in this view of life** . . . . Here **grandeur**, which relates to B, is unusual in being at the same time both Theme and New; hence it is doubly prominent.[17] On a first reading, **in this view of life** seems to complete the clause; and since it is anaphorically cohesive (by reference **this**, and by lexical repetition of **life**) it is read as not only Rheme but also Given. It then turns out that Darwin has misled us with a grammatical pun, and that **life** actually begins a new clause, **life . . . having been originally breathed into a few forms or into one**, all of which is a projected Qualifier to **view** ('the view that life was originally breathed . . .'). This makes **life** thematic, and (since no longer anaphoric) a continuation of the motif of life forms in II (2.); this motif is then carried on into the New, in **(breathed) into a few forms or into one**. The next clause turns out to be another projected Qualifier to **view**, paratactically related to the last yet finite where the other was non-finite; furthermore it is a hypotactic clause complex in which the dependent clause comes first. The dependent clause has as Theme **this planet**, relating cohesively to **the world** at the very beginning of II (1.): and as New **(has gone cycling on) according to the fixed law of gravity**, where **law of gravity** derives from motif C (natural laws) but shifts the attention from the temporally organized world of biology to the timeless universe of physics. The final clause, the culmination of the projected 'view' in which there is grandeur, has the Theme from III (**from so simple a beginning**, with anaphoric **so**); the Rheme takes up the motifs of II, **endless forms**, and B (3), **most beautiful and most wonderful**, leading to the final New element, the verbal groups **have been, and are being, evolved**.

   This resounding lexicogrammatical cadence brings the clause, the sentence, the paragraph, the chapter and the book to a crashing conclusion with a momentum to which I can think of no parallel elsewhere in literature — perhaps only Beethoven has produced comparable effects, and that in another medium altogether. Phonologically, the coordination of **have been, and are being**, forces a break in the rhythm (further reinforced by the surrounding commas) that directs maximum bodyweight on to the final word **evolved**. Grammatically, the word **evolved** has to resolve the expectation set up by the ellipsis in the uncompleted verbal group **have been**. Semantically, **evolved** has to resolve the conflict between **so simple a beginning** and **endless forms most beautiful and most wonderful**. All that is only what the word is expected to achieve within its own clause. In addition, within the projected clause complex, it has to complete the complex proportion between physical and biological processes:

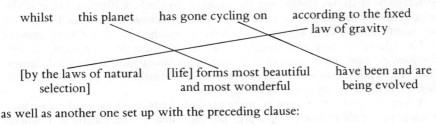

as well as another one set up with the preceding clause:

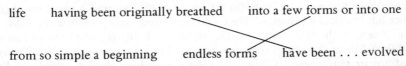

Within the sentence, the word **evolved** has to carry a culminative prominence to match the initiating prominence carried by **grandeur** (as Theme/New) at the beginning. Within these two paragraphs, it has to pick up the thematic motif of explanation, and to secure total commitment to one explanation and rejection of the other. It is here that the selection of voice becomes important: since the verbal group is passive, the responsibility for evolution is clearly lodged with the Creator (there is an external agency at hand; it is not . . . **have been, and are, evolving**). Yet all this load of work is hardly worth mentioning beside the major responsibility the word **evolved** has to bear, along with the verbal group of which it is a part: that of sustaining the climax of 450 pages of intense scientific argument. This is the culmination towards which the entire text has been building up. It would be hard to find anywhere in English a sentence, or a clause, or a group, or a word that has been made to carry such an awesome semiotic load.

I do not know how long it took Darwin to compose these two paragraphs, or whether he reflected consciously on their construction as he was doing it — I imagine not. I certainly had no idea, when starting the analysis, of what I was going to find. I had the sense of a remarkable and powerful piece of writing, as the climax to a remarkable and powerful book; and it struck me that something of the effect of these two paragraphs might lie in the patterning of the Theme and of the New — that is, in the textual component within the grammar of the clause. It is important to stress that that is in fact all that I have been looking at in this paper; I have said almost nothing about cohesion or transitivity or mood or the clause complex or any of the other lexicogrammatical systems/processes that go into the makeup of a text. Some, at least, of these other features would undoubtedly show interesting and significant patterns if we were to analyse them with this or some comparable kind of functional grammatics.

It is pointless to try and classify a text such as this — to ask whether it 'is' a scientific treatise or a declaration of faith or an entertaining work of literature. It is a product of the impact between an intellectual giant and a moment in the space-time continuum of our culture, with all the complexity of meaning that that implies. With this very partial analysis — a fragment of the grammar of a fragment of the text — I have tried to suggest something of how this text takes its place in semohistory. Some of the thematic patterning here is like that which I described in the first part of the paper, which evolved primarily (I think) in the context of scientific endeavour; we can recognize instances where Darwin is backgrounding some point already covered, so getting it taken for granted, and moving on from it, by a logical-semantic 'process', to a foregrounded next stage; e.g., **from the war of nature . . . the production of the higher animals directly follows**. (There are more of this type in the more strictly 'scientific' passages; for example the account of the honeycomb in Chapter 7, pp. 255–6.) But the pattern has rather a different value here from that which it typically has in the context in which it evolved; Darwin's strategy is that of accumulating masses of evidence rather than moving forward logically one step at a time. And particularly at critical moments he moves into a more monumental mode, that of a writer producing a text which he knows is unique and will have a unique place in the history of ideas. What is important is that we should be able to use the same theory and method of linguistic analysis — the same 'grammatics' — whatever kind of text (or sub-text) we are trying to interpret, whether Tennyson or Darwin, Mother Goose or the *Scientific American*. Otherwise, if we simply approach each text with an ad hoc do-it-yourself kit of private commentary, we have no way of explaining their similarities and their differences — the aesthetic and functional values that differentiate one text from another, or one voice from another within the frontiers of the same text.

Notes

1   That is, there are many 'disjunctions'; see Lemke, (1984) especially pp. 132 Dialectal disjunctions are mainly phonetic; see the Prague school's concept of functional equilibrium in phonology.
2   The term 'diatypic' is taken from Gregory (1967). The term 'register' was first used in this sense by Reid (1956); see Halliday, McIntosh and Strevens, (1964).
3   The concept of register should therefore be defined so as to make explicit the dimension of power, as pointed out by Kress (1988a), and Fairclough, (1988).
4   For this and other aspects of the systemic-functional grammar referred to throughout this paper see Halliday (1985a).
5   The Given + New structure is not, in fact, a structure of the clause; it constructs a separate unit (the 'information unit') realized by intonation as a tone group. In spoken English the typical (unmarked) discourse pattern is that where one information unit is mapped on the one clause; further semantic contrasts are then created by departure from this unmarked mapping. In written English there are of course no direct signals of the information unit; while the unmarked mapping may be taken as the typical pattern, a great deal of systematic variation will show up if the text is read aloud.
6   This is, obviously, a very sketchy and selective account. See Halliday (1985c) chapters 7 and 9; and Table 9(3), pp. 306–7.
7   For this distinction see Halliday and Hasan, (1976); chapter 5, 'Conjunction', especially pp. 240–4. Here we refer to "external' (*in rebus*) and 'internal' (*in verbis*) conjunctive relations.
8   We do not of course transcend these oppositions; the nearest we get to a position of neutrality, in the sense of being able to accommodate the complementarities on a higher stratum, is in the discourse of mathematics and of linguistics — as thematic rather than disciplinary discourses (perhaps now computer science and semiotics).
9   See Fries (1981/1983). For a more recent discussion see the same author's 'Toward a discussion of the flow of information in a written English text', in William C. Mann and Sandra Thompson (Eds) (1992), *Discourse Description: diverse analyses of a fund-raising text*, Amsterdam: Benjamins.
10   The boundary between Given and New is in any case fairly indeterminate. What is clearly marked by the intonation contour is the information focus: that is, the culmination of the New (signalled by tonic prominence). There is some prosodic indication where the New element begins, but it is much less clear (hence the move from Given to New is often regarded as continuous). See also note 15 below.
11   Ranking clauses are those which are not embedded (rankshifted); they enter as clauses (either alone, or in paratactic or hypotactic relation with others) into clause complexes (sentences). Embedded clauses are not considered, because they do not enter into clause complexes but function inside the structure of a nominal group, and present little choice of textual (thematic or informational) organization; thus their Theme — Rheme and Given — New structure has no significance for the overall patterning of the discourse.
12   Chapter 9, 'Imperfection of the geological record'.
13   In **to my mind** the 'me' has no role in the transitivity structure (no ideational function). In **when I view ...**, the same 'me' has been transformed into a

thinker, with a highly significant role in transitivity — as Senser in a mental process; note here also the lexical slip from **view** = 'observe', suggested by **when I view all beings**, to **view** = 'opine', a reinterpretation forced on the reader by the subsequent **as**.

14 Based on my own reading of the text: on the construction into information units and location of information focus.

15 This option is not available to a motif functioning as Theme, since (almost) all thematic elements are nominals (any clause functioning as Theme has first to be nominalized). Instances such as **transmit likeness, transmit progeny, have left no descendants**, etc. illustrate the point made in note 10 above; in my reading the New could be heard as beginning with the verb in each case. I have used the more cautious interpretation, restricting it in most instances to the final (culminative) element.

16 I have treated all these as falling within the New, rather than attempting to analyse them further; a list tends to have special rhythmic and tonal properties of its own.

17 That is, it clearly represents a 'marked' mapping of information structure on to thematic structure, characteristic of such existential clauses.

Chapter 6

Language and the Order of Nature*

M.A.K. Halliday

Order out of Language

Out of the buzz and the hum in which mankind has been evolving — itself a kind of conversation, to our present way of thinking — has emerged what Rulon Wells once called the 'distinctively human semiotic' : a special form of dialogue powered by a system we call language. With this we talk to each other; and in the process we construct the microcosmos in which each one of us lives, our little universes of doing and happening, and the people and the things that are involved therein.

And in the course of this semiotic activity, without really becoming aware of it we have also been construing the two macrocosmic orders of which we ourselves are a part : the social order, and the natural order. For most of human history, these deeper forms of dialogue have depended on substantially the same resource: ordinary, everyday, spontaneous, natural spoken language — with just some 'coefficient of weirdness' such as Malinowski (1935) found in the more esoteric contexts of its use.[1]

All this dialogic construction is, by definition, interactive. At the micro level, we get to know our fellow-creatures by talking to them and listening to them; and they respond to us in the same natural language. At the macro level, the 'dialogue with nature', brilliantly scripted by Prigogine and Stengers (1985) in their book *Order Out of Chaos*, is also interactive; but in another guise. When we want to exchange meanings with physical or biological nature we have to process information that is coded in very different ways, and that may need to go through two or three stages of translation before we can apprehend it.

We have always assumed that it *can* be translated; that the information coming in can in the last resort be represented and transmitted through the forms of our own natural languages. In fact, up until the last few millennia, no conceivable alternative could ever present itself; because language was beyond the range of our conscious reflection. It was simply part of ourselves — the label 'natural' language is entirely apt. Herbert Simon (1969), in his *Sciences of the Artificial*, classified language among the artificial phenomena; but he was wrong. Language is as much a product of evolution as we are ourselves; we did not manufacture it. It is an evolved system, not a designed system: not something separate from

* This chapter is taken from Nigel Fabb et al. (Eds), *The Linguistics of Writing*, Manchester University Press, 1987, chapter 9.

humanity, but an essential part of the condition of being human. These natural languages, then, sufficed to enable us to interpret both facets of our wider environment, the social order and the natural order; these were, after all, construed by generalizing and abstracting from the micro-environments in which language had evolved all along.

It is just within the last hundred generations or so that some element of design has come into natural language; and just in certain cultural-historical contexts, those in which language has come to be written down. Writing has been an inherent part of the process. In these contexts the dialogue with nature has begun to take on new forms; we have learnt to measure, and to experiment; and to accompany these new semiotic modes, our languages have spawned various metalanguages — the languages of mathematics and of science. These are extensions of natural languages, not totally new creations; and they remain in touch. Even mathematics, the most 'meta-' of all the variants of natural language, is kept tied to it by an interpretative interface — a level one metalanguage which enables mathematical expressions to be rendered in English, or Chinese, or other forms of distinctively human semiotic.

Does Language Cope?

Now and again some part of the dialogue breaks down, and then it becomes news — like London Bridge; as long as it stays up it is not news, but when it falls down it will be. Yet what is really newsworthy about language is how rarely it does fall down. The demands that we make on the system are quite colossal; how is it that it so seldom gets overloaded?

As far as the social order is concerned, we can watch language at work construing this from a child's earliest infancy, because from the moment of birth language intercedes, mediating in the dialogue between an infant and its care-givers. This kind of language, which is language for loving and caring rather than for knowing and thinking, would seem to have no great demands being made on it, and only in pathological cases is it likely to break down. Since here language is not being required to refer, its success is judged other than referentially. But this least referential kind of discourse is in fact actively enrolled in constructing the social order. Predictably, since the social order is highly complex, the language that is creating it is also highly complex. Only, the two complexities are not related in a straightforward referential fashion. Language creates society; but it does so without ever referring to the processes and the structures which it is creating (see Hasan, 1986).

This appears surprising only because we are obsessed with the referential properties of language. Yet language is 'not constrained by the need to refer':[2] even if a mother was aware of the features of the social order that her dialogue was bringing into being, her baby could scarcely be expected to understand them. She cannot talk to her child about social values, statuses and roles, decision-making hierarchies and the like. Yet all these are brought into being by her use of the grammar, as Hasan (forthcoming) has convincingly demonstrated; and this is why when people want to change the conditions of the dialogue, and the structures it is setting up, they do so by changing the grammar — thus illustrating how well the grammar is doing its job. The complaint is not that the language is not functioning properly, but that it is functioning all too well — it is the social order construed by it that is being objected to. But mostly the design for change is drawn up only at the surface of the language, rather than at the much less accessible, cryptotypic level of patterning by which the structures are really installed.

When we come to consider how language creates the *natural* order, we might expect matters to be different. This is an order of happenings, and things, and language is our primary means of reflecting on these. Here, presumably, the essential function of language is to refer — to make contact with what is 'out there'. But this too we shall have to call into question.

In this sphere there have been from time to time complaints of a different kind. It is often objected that language is letting us down; and this especially at certain times in history, when the pace of the dialogue is quickening and knowledge is accumulating very fast. At such times there arise proposals for improving language, making it a more effective tool for recording and extending our knowledge.

Thus the precursors and contemporaries of Newton set about remodelling language: first simply as shorthand, then as universal character (a script that would be the same for all languages, based on the logographic principle of Chinese), then as 'real character' (a new universal written language, with its own realizations in speech — the famous systems of Cave Beck, Dalgarno and John Wilkins), and finally conceived of as a calculus, a semiotic not so much for recording and transmitting knowledge as for creating it — a tool designed for thinking with, like the mathematical calculus newly invented by Leibniz and by Newton himself. (This last was never realized; nor were any of the earlier systems that were realized ever used.)

Since 1900 the call has been heard again, heralded perhaps by de Broglie's famous observation that 'physics is in suspense because we do not have the words or the images that are essential to us'. We can follow this motif through Einstein, Bohr and Heisenberg, down to the present day as it becomes increasingly specific.[3] David Bohm devotes a whole chapter to language, in which he objects that 'language divides things into separate entities', and so distorts the reality of 'undivided wholeness in flowing movement' ; and he proposes a new form of language called the 'rheomode', which gives the basic role to the verb rather than the noun.[4]

We shall return to these objections in a moment. But despite the shortcomings which the natural scientists of these two periods have found in the languages they had to work with, science has continued to progress — and to change direction fundamentally on both occasions. Languages have not given way beneath its weight, nor are there any very obvious signs of overload. By and large, the dialogue has worked. One thing that Bohm and his predecessors may have overlooked is that you do not need to keep engineering a language in order to change it; it will change anyway — because that is the only way it can persist.

Language as Dynamic Open System

Every language is constantly renewing itself, changing in resonance with changes in its environment. But this is not an incidental fact about language; it is a condition of its existence as a system — and without language as system there could be no dialogue at all.

The earliest linguists of India, Greece and China all recognized that languages change in their expression — in their phonetics and morphology. These effects have now been shown to be statistical.[5] variation sets in, from a variety of sources, internal and external; and this variation can either become stabilized, so that the system becomes inherently variable at that point, or a 'variable rule becomes categorical' and we say that sound change has taken place. The terminology ('variable rule', and so on) is unfortunate, since it leads people — or perhaps confirms their inclination — to look for hidden variables, so that all variable rules can be reduced to categorical status. But the data resist such interpretations; it is simply not the case that if we knew everything there was to know then we could predict every instance. In other words, variability in language is not a limitation of the observer; it is a feature of the system, and hence the statistically defined behaviour of the micro particles of language — for example the realization of a particular vowel as a fronted or a backed variant — can induce the system to change.

Such expression variables are alternative realizations of some higher level constant (e.g., 'the phoneme /a/'), which therefore constitutes the entry condition to that particular little system. To understand changes in the *meaning potential* of language we need to consider analogous statistical effects on the content plane (see Nesbitt & Plum, 1988). Consider a grammatical system such as past /present/ future primary tense in English, interpretable semantically as deictic reference to a linear scale of time (the traditional description was 'time relative to the moment of speaking'). In any instance of the context which serves as entry condition to that system — in grammatical terms, any instance of a finite clause — each term has an inherent probability of occurring. A speaker of English 'knows' these probabilities; having heard, by age 5, say, about half a million instances he has a statistical profile of the lexicogrammar of the language. It is by the same token (more specifically, by the same set of tokens) that he knows the lexical probabilities: that **go** is more frequent than **grow** and **grow** is more frequent than **glow**. Now, grammar and lexis are simply the same phenomenon looked at from two different ends; but one difference between them is that the patterns we treat as grammatical are those which are buried much deeper below the level of people's consciousness, and so these patterns, and the probabilities associated with them, are much harder for people to become aware of — many people reject grammatical probabilities when they are told about them: they feel insulted, and take them as affronts to the freedom of the individual. Lexical patterns are nearer the surface of consciousness : hence lexical probabilities are quite readily recoverable, as a sense of 'this word is more frequent than that', and are therefore found easier to accept. (Logically, of course, there is no reason why being told that one is going to use **go** more frequently than **glow** should be any less threatening than being told one is going to use past more frequently than future; but these observations tend to provoke very different responses.)

Now, just as, when I listen to the weather report every morning, and I hear something like 'last night's minimum was six degrees, that's three degrees below average', I know that that instance has itself become part of, and so altered, the probability of the minimum temperature for that particular night in the year — so every instance of a primary tense in English discourse alters the relative probabilities of the terms that make up the primary-tense system. Of course, to make these probabilities meaningful as a descriptive measure we have to sharpen the focus, by setting conditions: we are not usually interested in the average temperature at all times anywhere on the surface of the globe (though this is a relevant concept for certain purposes), but rather in the probable daily minimum on Sydney Harbour at the time of the winter solstice. There are various dimensions conditioning grammatical probabilities: we might specify the context of situation — for example the discourse of weather forecasting, which will considerably increase the weighting for future tense;[6] or we might specify a number of other concurrent grammatical features, such as whether the clause is declarative or interrogative. (There are also the transitional probabilities of the text as a stochastic process.) The more local the context, of course, the greater the moderating effect of a single instance.

Lemke has pointed out that many human systems, including all social-semiotic systems, are of a particular kind known as 'dynamic open systems'.[7] Dynamic open systems have the property that they are metastable: that is, they persist only through constant change; and this change takes place through interactive exchanges with their environment. In the course of such interaction, the system exports disorder; and in the process of exporting disorder, and so increasing the entropy of its environment, the system renews itself, gains information, imports or rather creates order and in this way continues to function. The system exists only because it is open. But it is now no longer itself; for such a system, the state of being is one of constant becoming. Language — natural language — is certainly a system of this general type.

Language (like other social semiotic systems) is a dynamic open system that achieves metastability through these statistical processes. Instances affect probabilities; from time to time probabilities thus rise to one or fall to zero, so that quantitative effects become qualitative and the system maintains itself by evolving, through a process of constant change.

In an ideal system, one having two states that are equiprobable, there is no redundancy. Once we depart from equiprobability, redundancy sets in. In all open systems the probabilities are skewed, so that the system carries redundancy. Lemke shows that a semiotic system is one that is characterized by redundancy *between* (*pairs of*) *its subsystems*: what he refers to as 'metaredundancy'. To illustrate from the classic Hjelmslevian example of the traffic lights: the system has certain states, red yellow and green. Since these are not equiprobable there is redundancy among them, at that level; but this simple redundancy is relevant only to the engineer who designs and instals them. There is then a *metaredundancy* between this system and the system of messages 'stop/go': this is the first order metaredundancy that defines the signifier and the signified. There is then a second order metaredundancy: this in turn 'metaredounds' with the system of behaviour of drivers approaching the signal: they stop, or else they drive on. And so on.

In other words: what the system 'says' (the wording: red/green), redounds with what it 'means' (the meaning: stop/go),which in turn redounds with what it does. But 'it' is a human system: it is people who drive the cars, people who construe the semantic opposition of stop/go, and people who switch on the lights, or at least programme the machine to do it for them. Traffic lights are in fact part of the social semiotic, even though I am using them here simply as an analogy for discourse that is 'worded' in the more usual sense — that is, in the form of lexicogrammar.

The Emergence of Metalanguages

Thus viewed as a social semiotic, language is a dynamic open system, probabilistic, and characterized by metaredundancy. These *n*-order meta-redundancies define the levels, or strata, of the system: the relationship of meta-redundancy is the general relationship whose manifestation in language we are accustomed to referring to as 'realization'. Such a system is good for thinking with and good for doing with, these being the two complementary facets of all human semiosis.

When either of these facets comes under pressure, the system responds by creating special varieties of itself to meet the new demands. So in a period of rapid growth of science and technology new metalanguages appear. These new forms of language both are created by and also create the new forms of knowledge — since what we call knowledge is simply a higher level of meaning, still linked to the grammar by the chain of metaredundancies. But it is at this point that the functioning of language starts to become problematic.

Let me refer again to David Bohm's *Wholeness and the Implicate Order*. Bohm is dissatisfied with the way language (as he sees it) fails to meet the demands of the new dialogue with nature, and he proposes the 'rheomode' — a form of language that would represent the flux of things, and construe experience as dynamic rather than static. His suggestions are simplistic and confined to a few variations in derivational morphology: for example, to get away from 'a language structure in which nouns are taken as basic, e.g., "this notion is relevant"', we reinstate the verb **to levate**, meaning 'the spontaneous and unrestricted act of lifting into attention any context whatsoever, . . .'; we then introduce the verb to **re-levate**, 'to lift a certain context into attention again,' whence **irrelevation, levation** and so on. (Bohm, 1980, pp. 34) But the motive is clear: a new language is needed to encode a new view of reality.

There have been frequent assertions, throughout the history of quantum mechanics and the physics that derived from it, that it is impossible to talk about quantum ideas in language as it was received. The language of physics is under stress; and some of the more farfetched notions such as the 'many worlds' interpretation proposed by Everett and Wheeler — that there are as many alternative realities as there are quantum events — might be used to illustrate the incapacity of our natural language-based metalanguages to cope with these new semiotic demands. The metalanguages are too determinate, too rigid, too unable to accommodate complementarities. They cannot tell us 'that *all* is an unbroken and undivided whole movement, and that each "thing" is abstracted only as a relatively invariant side or aspect of this movement'. (Bohm, 1980, p. 47)

Before examining these charges further, let us note an interesting paradox. When logicians and philosophers complain about language, their usual complaint is that it is too vague. When scientists find language letting them down, it is generally because it is too precise, too determinate. In part, no doubt, this reflects their two different ideologies. For the logician, if two things conflict they cannot both be true — so language should force them to reject the one or the other; and if language does not do this, then it is too loose, too vague. For the scientist, on the other hand, if two things are both true they cannot conflict — so the language should help them to accommodate both, and if it doesn't, it is too rigid, too determinate. (What I am labelling 'the logician', 'the scientist', are of course two different ideologies — not the individual members of these two repected professions.)

But they are also probably talking about different languages. The logicians are thinking of non-technical natural language, from which their artificial languages, including mathematics, were first derived. The scientists are thinking of their own technical metalanguages that have been constructed on the basis of natural language: the various registers of physics, for example. And these scientific metalanguages are among the more designed varieties of human language — hence, like all designed systems, they do tend to be rigid and determinate. These are the very features which make such metalanguages unsuitable for just that purpose for which they were in fact designed: the dialogue with nature, for which it is essential to be able to mean in terms that are dynamic, non-compartmental and fluid — and above all, that do not foreclose.

The irony is, that is exactly what natural language is like: dynamic, non-compartmental and fluid. But it has got smothered under the weight of the metalanguages that were built upon it.

Levels of Consciousness in Language

Let me begin this section with a quotation from Prigogine and Stengers' book *Order Out of Chaos* that I referred to at the outset:

> [In quantum mechanics] there is an irreducible multiplicity of representation for a system, each connected with a determined (i.e., decided upon by the investigator — MAKH) set of operators. This implies a departure from the classical notion of objectivity, since in the classical view the only 'objective' description is the complete description of *the system as it is*, independent of the choice of how it is observed. . . .
>
> The physicist has to choose his language, to choose the macroscopic experimental device. Bohr expressed this idea through the principle of complementarity, . . .
>
> The real lesson to be learned from the principle of complementarity, a lesson that can perhaps be transferred to other fields of knowledge, consists in emphasizing the wealth of reality, which overflows any single language, any single logical structure. Each language can express only part of reality. [Prigogine and Stengers, 1984, p. 225]

Here Prigogine and Stengers are of course talking about 'languages' in the sense of conceptual constructs; and they go on to say:

> No single theoretical language articulating the variables to which a well-defined value can be attributed can exhaust the physical content of a system. Various possible languages and points of view about the system may be complementary. They all deal with the same reality, but it is impossible to reduce them to one single description.

It is my contention that natural language — not as it is dressed up in the form of a scientific metalanguage, but in its commonsense, everyday, spontaneous spoken form — does in fact 'represent reality' in terms of complementarities; and that these are complementary perspectives in precisely the sense in which Bohr was using the term.[8] Only, it does so non-referentially. Just as language construes the social order without referring to the system it is constructing, so likewise language construes the natural order — through the unconscious, cryptotypic patterns in the grammar, which create their own order of reality independently of whatever it is they may be being used to describe.

I shall illustrate the complementarities inherent in this 'de-automatized' sphere of the grammar in just a moment. But first we must recognize a problem. The features I am referring to in natural language are features of the 'cryptogrammar'; they function way below the usual level of consciousness. And the problem is, that when we start reflecting on them, bringing them up to our conscious attention, we destroy them. The act of reflecting on language transforms it into something alien, something different from itself — something determinate and closed. There are uses for closed, determinate metalanguages; but they can represent only one point of view about a system. The language of daily life, which shapes our unconscious understanding of ourselves and our environment, is a language of complementarities, a rheomode — a dynamic open system. The question is whether we can learn to use it to think with consciously. It may be impossible. I don't mean that it is impossible to *understand* the cryptogrammar of a natural language, but that its reality-generating power may be incompatible with explicit logical reasoning.

I have tried out a simple strategy for exploring the more unconscious features of the grammar. I selected a text — the headlines of a news broadcast, which I had taken down verbatim from the radio; I read it aloud to a group of students, and asked them to recall it. They gave me the motifs: death, disaster, violence and the like. I pressed them further: what was actually said? This time they gave me words: a list of the lexical items used, recalled with considerable accuracy although most of them had not figured in their first responses. Let me call the motifs level zero, and the lexical responses level one. I pressed them for a more specific account (still without reading the passage again), and they gave me the more exposed parts of the grammar: the word, group and phrase classes, the derivational morphology and so on. This exposed grammar we will call level two. I pressed them once more; and this time — since they were students of linguistics — they began to get to level three, the hidden grammar (the cryptotypes, in Whorf's terminology): the transitivity patterns, the grammatical metaphors and so on.

In our normal everyday concern with language we simply attend to the motifs. We are not concerned with wordings, and do not trouble even to remember them. We behave as if the metaredundancy — the realization of meanings in lexicogrammar — is simply an automatic coding. If asked to reflect on the wording, we focus on the lexical end of the spectrum: the words, or rather the lexical items — since this is the edge that is nearest the domain of conscious attention. It takes much more effort to attend to the more strictly grammatical zone, especially to its more cryptotypic regions. And when we get there, we find ourselves back at the motifs again; but this time with a greatly heightened understanding, because now we can see why the text meant what it did, and we can appreciate the deeper ideological content of the discourse — the messages we had received without becoming aware of them.

The process of reflecting on natural language can be modelled in terms of these four levels of consciousness:

'meaning'	(semantic level)	level 0: 'motifs'
		⎧ level 1: 'words'
'wording'	(lexicogrammatical level)	⎨ level 2: 'phenotypes'
		⎩ level 3: 'cryptotypes'

— where the spiral (cryptotypes as hidden motifs) in turn represents the dialectic of metaredundancy. Or, to put this in more familiar semiotic terms: the signified constructs the signifier (by 'realization' — grammar in its automatized function), and the signifier constructs the signified (grammar, especially the cryptogrammar, in its de-automatized function). The problem of turning the cryptogrammar of a natural language into a metalanguage for reasoning with is that it has to become automatized — that is, the grammar has to be made to describe, instead of constructing reality by not describing, which is what it does best.

Everyday Language as a Theory of the Natural Order

I will try to enumerate some features of natural language, as embodied in our everyday informal discourse from earliest childhood, that constitute for us a theory of reality. They are features common to all languages, but in respect of which each language presents its own particular mix; I make one or two references to English, but in the main they are set out in general terms that could be applied to all.

- CLAUSAL STRUCTURES: the organization of meanings in lexico-grammatical form (as 'wordings'). The gateway through which meanings are brought together and realized in ordinary grammar is the clause; and the clause nucleus is a happening (Process + Medium, in systemic terms). So natural languages represent reality as what happens, not as what exists; things are defined as contingencies of the flow.
- PROJECTION: the general relation underlying what grammarians call 'direct and indirect speech'. The system of projection construes the whole of experience into two different kinds of event: semiotic events, and other events; the latter can then be transformed into semiotic events by processes of consciousness.
- EXPANSION: logical-semantic relationships between events. Two events provided they are of the same kind (as defined by projection) may be related to one another by one of a set of logical-semantic relations, such that the second one defines, extends, or in some way (such as time or cause) correlates with the first.
- TRANSITIVITY: the theory of processes (1). Natural languages construe experience out of different types of process; this plurality is universal, though the details of the system vary. English sets up 'outer' processes, those of the world perceived as external; 'inner' processes, those of (human-like) consciousness; and processes of attribution and representation. All are distinguished in the cryptogrammar.

- TRANSITIVITY: the theory of processes (2). With regard to (at least) the 'outer' processes, natural languages incorporate two models : the transitive, which interprets 'mechanically', in terms of transmission, and the ergative, which interprets 'scientifically', in terms of causation. These two are complementary; the generalizations they make contradict each other, but every clause has to be interpreted as both.
- TENSE and ASPECT: the theory of time. Similarly, natural languages embody two models of time: a theory of linear, irreversible time, out of past via present into future (tense), and a theory of simultaneity, with the opposition between being and becoming, or manifested and manifesting (aspect). Languages have very different mixtures (English strongly foregrounds linear time); but probably every language enacts both, and again the two are complementary in the defined sense.

In these and other features of their 'hidden' grammars, ordinary languages in their everyday, commonsense contexts embody highly sophisticated interpretations of the natural order, rich in complementarities and thoroughly rheomodal in ways much deeper than Bohm was able to conceive of. To be more accurate, we would have to say that it is these features *in a system of this dynamic open kind* that construe reality for us in this way. The system itself must be a metastable, multi-level ('metaredundant') system — that is, a human semiotic — with the further property that it is 'metafunctional': it is committed to meaning more than one thing at once, so that every instance is at once both reflection and action — both interpreting the world and also changing it.

We have been constantly reminded of 'the impossibility of recovering a fixed and stable meaning from discourse'. Of course this is impossible; it would be a very impoverished theory of discourse that expected it. But it is entirely possible — as we all do — to recover from discourse a meaning of another kind, meaning that is complex and indeterminate. The reason it is hard to make this process explicit is that we can do so only by talking about grammar; and to do this we have to construct a theory of grammar: a 'grammatics', let us call it. But this 'grammatics' is itself a designed system, another scientific metalanguage, with terms like 'subject' and 'agent' and 'conditional' — terms which become reified in their turn, so that we then come to think of the grammar itself (the real grammar) as feeble and crude because it doesn't match up to the categories we've invented for describing it. But of course it's the grammatics — the metalanguage — that is feeble and crude, not the grammar. To borrow Whorf's famous simile, the grammatics (grammar as metalanguage) is to the grammar (the language) as a bludgeon is to a rapier — except that a better analogy might be with the hand that wields the rapier. If the human mind can achieve this remarkable combination of incisive penetration and positive indeterminacy, then we can hardly deny these same properties to human language, since language is the very system by which they are developed, stored and powered.

The Need for Plurality of Language

To quote Prigogine and Stengers again: 'Whatever we call reality, it is revealed to us only through active construction in which we participate' (1985. p. 293). But, as they have already told us, 'the wealth of reality . . . overflows any single language, any single logical structure. Each language can express only part of reality'.

I have suggested that our natural languages do possess the qualities needed for interpreting the world very much as our modern physicists see it. But from the time when our dialogue with nature became a conscious exercise in understanding, we have come to need more than one grammar — more than one version of language as a theory of experience. Rather, we have needed a continuum of grammars, from the rheomodal pole at one end to something more fixed and constructible at the other. For our active construction of reality we had to be able to adopt either a dynamic, 'in flux' perspective or a synoptic, 'in place' perspective — or some mixture of the two, with a complementarity between them.

So our language began to stretch, beginning — as far as the West is concerned — with the explosion of process nouns in scientific Greek from B.C. 550 onwards (e.g., *kinesis* 'movement', from *kineo* '(I) move'), and culminating (so far!) in the kind of semantic variation found in pairs such as:

experimental emphasis becomes concentrated in testing the generalizations and consequences derived from the theories	we now start experimenting mainly in order to test whether things happen regularly as we would expect if we were explaining in the right way
1-attic	1-doric

Let me label these two styles the 'attic' and the 'doric'. The attic mode is not of course confined to abstract scientific discourse; 2–attic is from a television magazine:

he also credits his former big size with much of his career success	he also believes that he succeeded in his career mainly because he used to be big
2-attic	2-doric

Represented in this new, 'attic' style, the world is a world of things, rather than one of happening; of product, rather than of process; of being rather than becoming. Whatever metaphor we use to label it with — and all these paired expressions capture some aspect of the difference — the emergence of the new attic forms of expression added a new dimension to human experience: where previously there had been one mode of interpretation, the dynamic, now there were two, the synoptic and the dynamic — or rather, two poles, with varying degrees of semantic space possible between them. There are now two ways of looking at one and the same set of phenomena.

The two are complementary, like wave and particle as complementary theories of light. Any aspect of reality can be interpreted either way; but, as with wave and particle, certain aspects will be better illuminated with the one perspective and others with the other. The doric style, that of everyday, commonsense discourse, is characterized by a high degree of grammatical intricacy — a choreographic type of complexity, as I have described it: (Halliday, 1986) it highlights processes, and the interdependence of one process on another. The attic style, that of emergent languages of science, displays a high degree of lexical density; its complexity is crystalline, and it highlights structures, and the interrelationships of their parts — including, in a critical further development, *conceptual* structures, the taxonomies that helped to turn knowledge into science.

There was thus a bifurcation in the metaredundancy pattern, leading to the duality of styles that Rulon Wells spoke about at the conference whose aftermath we are celebrating here (he referred to them as 'nominal and verbal styles', but the distinction is really that of nominal and clausal) (Wells, 1960). Between the doric, or clausal, style and the attic, or nominal, style is a complementarity that itself complements the various first-order complementarities that we have already seen to be present within the doric system. But this second-order complementarity is of a somewhat different kind. The two perspectives are not on equal terms. The dynamic mode is prior; it comes first.

The dynamic mode is phylogenetically prior; it evolved first, along with the human species, whereas it is only in the last few millennia that the synoptic mode has come into being. It is also ontogenetically prior; it is what we learn as children and carry with us throughout life. Whenever we are speaking casually and unselfconsciously, in typically human dialogic contexts, we go on exploiting the dynamic mode, which as we have seen embodies the deep experience of the species in cryptogrammatic form. The synoptic mode, on the other hand, embodies the more conscious reflection on the environment that is stored in scientific knowledge; historically it is derived from the dynamic by the processes of grammatical metaphor. Of course, once in existence it can enter daily life; there is nothing very abstruse or formal about **every previous visit had left me with a feeling of discomfort . . .** Nevertheless it is a metaphoric derivative; the agnate **whenever I'd visited before I'd ended up feeling uncomfortable . . .** is a prior form of semiosis. So how does the more synoptic mode, the attic style arise?

Thanks to the metaredundacy principle, it is possible to introduce variation at any one level of language without thereby disturbing the patterning at other levels of the system (that is, without catastrophic perturbations; the consequences are seen in continued gradual changes such as I described earlier). It is even possible to replace an entire level of the system in this way; and this is what happened with the development of writing. Writing provided a new mode of expression — which could 'realize' the pre-existing content patterns without disrupting them. At the same time, it provided a new interface, another kind of instantiation through which changes in the system could take place.

Writing evolved in the immediate context of the need for documentation and recording. But it opened the way to an alternative theory of reality.

The Effects of Writing

Conditions arise in history — essentially those of settlement — where experience has to be recorded: we need to store knowledge, and put it on file. So we invent a filing system for language, reducing it to writing. The effect of this is to anchor language to a shallower level of consciousness. For the first time, language comes to be made of constituents — sentences — instead of the dependency patterns — clause complexes — of the spoken mode. And with constituency comes a different form of the interpretation of experience.

It is important not to oversimplify the argument at this point. Both language itself, and the dimensions of experience that are given form by language, are extremely complex; and instead of hoping to gain in popularity ratings by pretending all is simple we do well to admit the complexity and try to accommodate it in our thinking. Let me take just three steps at this point.

Writing brings language to consciousness; and in the same process it changes its semiotic mode from the dynamic to the synoptic: from flow to stasis, from choreographic to crystalline, from syntactic intricacy to lexical density. Note that this is *not* saying that writing imposes organization on language. On the contrary: there is every bit as much organization in spoken language as there is in written, only it is organization of a different kind. Written language is corpuscular and gains power by its density, whereas spoken language is wave-like and gains power by its intricacy. I am not, of course, talking about writing in the sense of orthography, contrasting with phonology as medium of expression; but about *written language* — the forms of discourse that arise as a result of this change of medium (by a complex historical process that is based partly on the nature of the medium itself and partly on its functions in society). Similarly in talking about spoken language I mean the forms of discourse which evolved over the long history of language in its spoken mode; the mode in which language itself evolved.

Writing puts language in chains; it freezes it, so that it becomes a *thing* to be reflected on. Hence it changes the ways that language is used for meaning with. Writing deprives language of the power to intuit, to make indefinitely many connections in different directions at once, to explore (by tolerating them) contradictions, to represent experience as fluid and indeterminate. It is therefore destructive of one fundamental human potential: to think on your toes, as we put it.

But, secondly, in destroying this potential it creates another one: that of structuring, categorizing, disciplinizing. It creates a new kind of knowledge: scientific knowledge; and a new way of learning, called education. Thus writing changed the social semiotic on two levels. Superficially, it created documentation — the filing of experience, the potential to 'look things up'. More fundamentally, it offered a new perspective on experience: the synoptic one, with its definitions, taxonomies and constructions. The world of written language is a nominalized world, with a high lexical density and packed grammatical metaphors. It is these features that enable discourse to become technical; as Martin has shown, technicality in language depends on, not writing as such, but the kind of organization of meaning that writing brings with it (see Martin, 1986b). Until information can be organized and packaged in this way — so that only the initiate understands it — knowledge cannot accumulate, since there is no way one discourse can start where other ones left off. When I can say

> the random fluctuations in the spin components of one of the two particles

I am packaging the knowledge that has developed over a long series of preceding arguments and presenting it as 'to be taken for granted — now we can proceed to the next step'. If I cannot do this, but have to say every time that particles spin, that they spin in three dimensions, that a pair of particles can spin in association with one another, that each one of the pair fluctuates randomly as it is spinning, and so on, then it is clear that I will never get very far. I have to have an 'expert' grammar, the kind of grammar that is prepared to throw away experiential information, to take for granted the semantic relations by which the elements are related to one another, so that it can maximize textual information, the systematic development of the discourse as a causeway to further knowledge. That kind of grammar shuts the layman out.

It would take too long to demonstrate in detail how this written grammar works. Let me refer briefly to its two critical properties: nominalization, and grammatical metaphor. Most instances involve a combination of the two. For example:

> such an exercise had the potential for intrusions by the government into the legitimate privacy of non-government schools

Apart from **had**, the clause consists of two nominal groups: **such an exercise** and **the potential for intrusions by the government into the legitimate privacy of non-government schools**. The second of these displays one of the principal devices for creating nominal structures : nominal group **non-government schools** embedded inside prepositional phrase **of non-government schools** embedded inside nominal group **the legitimate privacy of non-government schools** embedded inside prepositional phrase **into the legitimate privacy of non-government schools**; another prepositional phrase **by the government**; the two both embedded in the nominal group **intrusions by (a) into (b)**, itself embedded inside the prepositional phrase **for intrusions** ... , embedded inside the nominal group **the potential for**. ... And most of these embeddings involve grammatical metaphor: **potential**, nominal expression of modality 'be able to', perhaps even a caused modality 'make + be able to'; **intrusions**, nominal expression of process 'intrude'; **privacy**, nominal expression of quality 'private'; **legitimate**, adjectival expression of attitudinally qualified projection 'as they could reasonably expect to be', and so on. (That these are marked, metaphorical realizations in contrast to unmarked, 'congruent' ones is borne out in various ways: not only are the congruent forms developmentally prior — children typically learn to process grammatical metaphor only after age 8 or 9 — but also they are semantically explicit, so that the metaphorical ones can be derived from them but not the other way round. But note that the 'metaphor' is in the grammar; there is not necessarily any lexical shift.)

So to the third step. Writing and speaking, in this technical sense of written language and spoken language, are different grammars which therefore consitute different ways of knowing, such that any theory of knowlege, and of learning, must encompass both. Our understanding of the social and the natural order depends on both, and on the complementarity between the two as interpretations of experience. I sometimes ask teachers about this question: whether there are things in the curriculum they consider best learnt through talking and listening, and other things best learnt through reading and writing. They have seldom thought about this consciously; but their practice often reveals just such a complementarity — processes and process sequences, such as sets of instructions, and including logically ordered sequences of ongoing argument, are presented and explored in speech, whereas structures, definitions, taxonomies and summaries of preceding arguments are handled through writing. Thus the complementarity of speech and writing creates a complementarity in our ways of knowing and of learning; once we are both speakers and writers we have an added dimension to our experience.

Having proclaimed the complementarity, however, I shall now take a fourth step — and end up by privileging speech. Again I stress that I am not talking about the channel; we can all learn to talk in written language, and a few people can manage the harder task of writing in spoken. I am talking about the varieties — spoken language and written language — that arose in association with these two channels. So by speech I mean the natural, unself-monitored discourse of natural dialogue: low in grammatical metaphor, low in lexical density, high in grammatical intricacy, high in rheomodal dynamic. This is language as it evolved as a dynamic open system; these are the features that keep it open, in the far-from-equilibrium state in which it enacts, and so construes, the semiotic parameters of our social, biological and physical levels of being. The frontiers of knowledge, in a post-quantum *nouvelle alliance*,[9] need a grammar of this kind to map them into the realm of 'that which can be meant'. But this mapping does not depend on reference, with the grammar being used in an automatized way to describe. If quantum ideas seem inexpressible, this may be because we have tried too hard to express them. They are almost certainly there already; what we must learn to do is to think grammatically — to recognize the ideological interpretant that is built into language itself.

Linguistics as Metatheory

We have been saying that natural language is a theory of experience. But it is clear that language is also a *part* of human experience. Thus the system has to be able to include itself.

Lemke (1984) has pointed out that a social-semiotic system of this kind is not subject to the Godelian restriction on self-reference.[10] Such a system can include itself in what it is describing — because it is a theory of praxis, of practices which operate irreversibly, ordered in time. Thus a grammar can also be, at the same time, a theory of grammar. I do not pretend to understand the argument at that point. But it is clear that, since we are interpreting language as it functions to create the natural and social order, and since it is itself part of that order, it must include itself in the description. And if we insist that linguists (*inter alios!*) should reflect on their own praxis as linguists, it is not just because such reflexivity is fashionable these days, but because we have learnt from quantum mechanics that the observer is an essential component in the total picture.

At the same time, as linguists we have learnt to be aware of the dangers of naive scientism. We have heard at length — and with justice — of the superficial importation of Darwinian concepts into nineteenth century historical comparatism; and there is a danger in the present situation too. Because we can see in post-quantum and far-from-equilibrium physics exactly the intellectual environment that is needed to make sense of language as we — independently — know it to be, we may all too easily latch on to these ideas and misinterpret them. And I am not claiming that, just by being aware of the danger, I have therefore avoided it myself.

Yet there are still two points to be made. One is that, for all its deflections and superficial applications where it did not fit, the Darwinian perspective was a fundamental one. Language has to be understood in a historico-evolutionary context, as part of evolutionary processes; the mistake is to apply these notions at places that are far too concrete and specific, instead of seeing them as the essential interpretative framework for our endeavours. And the same goes this time round, when all that comes from the sciences of nature resonates so sweetly with everything we as linguists have learnt to expect.

And secondly, it is not, in fact, the same scenario as before. History does not repeat itself — or only on the surface of things. This time, the communication is going both ways. We have become accustomed to accepting the privileged position of the natural sciences, which got their act together first: the nature of a scientific fact, notions of evidence and experimentation, and above all the relationship of the *instance* to the general principle — these were established first in physics, then in biology, and only a very late third in the social sciences.[11] So it was natural that physics should become the model for all the others.

Now, however, there are signs of a reversal. I quote David Bohm (1980, p. 123) again: 'the speed of light is taken not as a possible speed of an object, but rather as the maximum speed of propagation of a signal'. In the quantum world, events are explained not in terms of causality but in terms of communication, the exchange of information. This is what used to be called 'action at a distance' (Gribbin, 1985, p. 182). And from Prigogine and Stengers (1985, p. 14) once more:

> A new type of order has appeared. We can speak of a new coherence, of a mechanism of 'communication' among molecules. But this type of communication can arise only in far-from-equilibrium conditions . . .
>
> What seems certain is that these far-from-equilibrium phenomena illustrate an essential and unexpected property of matter: physics may henceforth describe structures as adapted to outside conditions . . . To use somewhat anthropomorphic language: in equilibrium matter is 'blind', but in far-from-equilibrium conditions it begins to be able to perceive, to 'take into account,' in its way of functioning, differences in the external world . . .
>
> The analogy with social phenomena, even with history, is inescapable.

All this points us in a new direction. From now on, the human sciences have to assume at least an equal responsibility in establishing the foundations of knowledge. Their coat-tailing days are over. But if so, our practitioners will surely have to learn to behave responsibly, instead of squandering themselves in the wasteful struggle for originality in which everyone else must be deconstructed so that each can leave his (or her) mark. We have to learn to build on our predecessors and move forward, instead of constantly staying behind where they were in order to trample them underfoot.

More importantly, it means that we have to examine our basic concepts in the light of their more general relevance to the sciences of life and of nature.

And as soon as we begin to do this, one thing stands out: that, among the human sciences, it is linguistics that finds itself inescapably in the front line. Partly because its object, language, is more accessible than those of sociology and psychology: more readily problematized, and seen to be opaque, than other forms of human behaviour. But more because, if we are to take seriously the notion that the universe is made of information, then we shall need a science of information — and the science of information is linguistics.

Why linguistics, rather than information science as at present constituted? Because natural language is the one non-designed human communication system, on the basis of which all other, artificial systems are conceived. It is presumably not a coincidence that, as technology has moved from the steam engine to the computer, so scientific explanations have moved from causality (limited by the speed of light) to communication (limited by the entropy barrier) (Prigogine and Stengers, 1985, p. 295). My colleague Brian McCusker (1983, p. 239) observed that the universe was now 'one, whole, undivided and *conscious*', so that the science of sciences had to be psychology. I think he should have said 'one, whole, undivided and *communicative*'. The source of interpretation of the universe as a communication system, insofar as this can be brought within the constraints of our understanding, has to be sought in grammar — the grammar of natural language, since that is where our understanding is born, and that is the means whereby we act and reflect on ourselves and our environment. If there is to be a science of sciences in the twenty-first century it will have to include linguistics — as at least a partner, and perhaps the leading partner, in the next round of man's dialogue with nature.

Notes

1 See the section entitled 'An ethnographic theory of the magical word'.
2 Quoted from David Butt, *The Relationship between Theme and Lexicogrammar in the poetry of Wallace Stevens*, Macquarie University Ph.D. thesis, 1984.
3 See Heisenberg, 1958. Especially Chapter X, 'Language and reality in modern physics'.
4 Bohm 1980. See especially Chapter 2, 'The rheomode — an experiment with language and thought'.
5 See Sankoff and Laberge, 1978. For the study of linguistic variation from this viewpoint, see Horvath, 1985. Especially Chapter 5, 'Analytical methodology'.
6 Based on an analysis of weather reports in the *New York Times* and *Chicago Tribune*, May-June 1985. See Halliday and Matthiessen, *Grammatical Metaphor in Text Generation* (forthcoming).
7 References throughout this paper are to J.L. Lemke's (1984) three articles 'Towards a model of the instructional process', 'The formal analysis of instruction' and 'Action, context and meaning'.
8 See Briggs and Peat, 1985 p. 54: 'Bohr approved of the uncertainty principle itself, believing it was an aspect of a deeper idea he called *complementarity*'. Complementarity meant the universe can never be described in a single, clear picture but must be apprehended through overlapping, complementary and sometimes paradoxical views. Bohr found echoes of this idea in classical Chinese philosophy and the theories of modern psychology.' He would have found them also in the grammar of natural languages.

9   The title of the original French version of Prigogine and Stengers' *Order Out of Chaos* was *La nouvelle alliance*.
10   See section 2.1 of 'Action, context and meaning', entitled 'Recursivity and praxis', pp. 71–73.
11   Culler identifies Durkheim, Freud and Saussure as those primarily responsible. See Culler, 1977.

Chapter 7

The Analysis of Scientific Texts in English and Chinese*

M.A.K. Halliday

The Language of Science

There are practical reasons for analyzing scientific texts. The most obvious is educational: students of all ages may find them hard to read, and we know from various research reports that, in English at least, the difficulty is largely a linguistic one. So if we want to do something about it we need to understand how the language of these texts is organized. Of course, if a text is hard to read the difficulty is bound to be in some sense linguistic, since texts consist entirely of language: but in the case of scientific writing it seems that there are certain features of the way meanings are organized, and the way they are 'worded', that present special problems for a learner, over and above the unfamiliar subject matter and its remoteness from everyday experience.

So people recognize that there is such a thing as 'scientific language', at least in the written mode; therefore it must be possible, using the theories and methods of linguistics, to say what its special features are. This raises two questions. One: *where* (at what level, or levels) in the linguistic system do we explain them? Are they in the lexicogrammar — that is, in the meanings constructed into sentences by the syntax and the vocabulary? in the discourse — the composition of the text and its rhetorical structure? or in some higher order — the ideological framework of knowledge, beliefs and value systems that form the cultural context of the text? All of these, of course, find their expression in what we call 'the words on the page'; but they involve different processes and conditions of interpretation. The answer may be, of course, that the concept of 'scientific language' involves all three.

The second question that arises is whether the phenomenon of scientific language is universal. That is to say, does each language have its own scientific register(s), so that we recognize scientific English (a special kind of English), scientific Chinese, and so on? or is there in some sense a single 'language of science' that is essentially the same no matter what language it is manifested in? Of course English and Chinese use different expression systems and different

* This chapter was originally presented at the International Conference on Research in Texts and Language, Xi'an Jiaotong University, in March 1989.

sentence grammars. But it might be that their scientific registers would have many features in common. If so, this makes the task of translation (including machine translation) much easier, since the transfer can take place at the lower levels of the linguistic system.

This paper mainly focuses on the second question, taking English and Chinese as the languages to be explored. Note that if we find (as of course we do) that there are some features in common to scientific English and scientific Chinese, this does not by itself tell us why. It might be that one is borrowed from the other: that in the twentieth century, scientific Chinese has been created by a process of transference (facilitated by actual translations) from English and other European languages. This is what has happened — or is commonly said to have happened — in many parts of the world where new scientific and technical forms of discourse have been created, sometimes in languages that were not previously written down. The other possibility is that the two evolved quite separately, but because of the nature of science, and scientific knowledge, they have grown to be very much alike. The truth may well be some mixture of the two (and either way we can compare them). But to the extent that shared features can be shown to be motivated, as opposed to conventional, to that extent we would expect them to evolve on their own, quite independently, without having to be transmitted by linguistic borrowing.

But some decision has to be taken regarding *where* to move in, in terms of the first of our questions above: do we investigate the lexicogrammar ('sentence grammar'), the text structure, or some higher order abstraction? Here I must take up an explicit theoretical position, since that will determine what the options are. I shall assume that *all* the features of a text participate in the creation of meaning; thus, whatever is characteristic of scientific writing *realizes* (dialectically, both determines and is determined by) the structure of scientific knowledge and its attendant ideological formations. But within this overall semogenic system-and-process, certain components are more accessible, more under conscious direction than others. To give an example from science, editors of scientific journals often give their authors specific instructions about how to write a scientific paper: introduce the topic, present the current state of knowledge, define a problem, describe the approach adopted to solving it, give the results accompanied by tables, figures, equations etc. where appropriate, draw conclusions, state what needs to be done next. This kind of rhetorical organization of the text, its structure as a piece of discourse, is easy to observe and fairly easy to prescribe. Furthermore it is minimally dependent on which language is being employed; such instructions can be carried out equally well in English, in Chinese, in Italian, Malay and so on. I am not saying that all scientific papers are written in the same format; on the contrary, there are considerable differences in the favoured construction of articles even in such culturally close languages as English, French and German (Clyne, 1991). But any such structure *could* be implemented in any language, and is readily borrowed from one language to another (thus translators will usually not alter the discourse structure of a text they are translating). It is clear that the evolution of such highly valued text types played a significant part in establishing Newtonian physics as the dominant paradigm for the natural sciences in renaissance Europe (Bazerman, 1988).

The text structure establishes the highly-valued form of scientific argumentation: what constitutes an acceptable canon of evidence, reasoning and proof. It models the 'macro' movements by which knowledge is extended and the scientific understanding progresses. By contrast, the lexicogrammar models the 'micro' movement, the smaller steps that, taken all together, comprise one episode in the macro-structure of the text. The lexicogrammar also models the conceptual structure, the taxonomies of categories that provide the scaffolding of scientific knowledge, based on the two fundamental semantic relationships of hyponymy (*a* is a *kind* of *x*) and meronymy (*b* is a *part* of *y*). And these components — the conceptual structure and the micro-argument — are much less accessible to observation and control, because they depend on linguistic patterns that speakers of a language are not aware of unless they engage in some technical linguistic analysis. By the same token, they are more closely tied to the semantic and lexicogrammatical structure of the particular language that is being used.

We will illustrate very sketchily some of the similarities and differences between English and Chinese in the way their lexicogrammar constructs scientific reality. We start from the lexical end — the vocabulary — because, within the lexicogrammar, it is the vocabulary, or lexicon, that is nearer the surface of consciousness. It is easier to pay attention to the vocabulary than to the grammar.

Taxonomies

All languages create taxonomies, based especially on hyponymy ('is a kind of'); and children begin to learn them from a very early age. For example: **fruit** is a kind of **food**; a **berry** is a kind of **fruit**; **blackberry**, **strawberry**, **raspberry** are kinds of **berry**. In the last stage in this example the language signals the taxonomic organization by including the word **berry** as a constituent morpheme of the word **blackberry** etc. English does not do this consistently; **carp**, **herring**, **cod/codfish**, **gemfish** are all equally kinds of **fish**. Chinese on the other hand does construct hyponymic sets consistently in this manner: every kind of fish ends in **yu** 鱼. Thus for a Chinese child the taxonomic structure of things in the environment is made very explicit; for an English child much less so. An English-speaking child does not learn cucumbers, melons, squashes, pumpkins and marrows as a related set; in Chinese they are all kinds of **guā** 瓜 (**dōngguā**, **xīguā**, **nánguā**, **wōguā**, **huángguā**, etc.). When it comes to constructing technical and scientific taxonomies, therefore, Chinese maintains the same principle — all gases are a kind of (i.e., compounded with) **qì** 气, all pointed instruments a kind of **zhēn** 针; and it extends also to abstract terms, so that **lièbiàn** 裂变 'fission' is a kind of **biàn** 变 'change', **tāishēng** 胎生 'viviparity' is a kind of **shēng** 生 'giving birth', **pínlù** 频率 'frequency' is a kind of **lù** 率 'rate', and so on. The English words **fission**, **viviparity**, **frequency** give no signal of their taxonomic status. The more highly technical terms in English are more likely to be constructed as taxonomic compounds; but if so, the components will be borrowed from Greek or Latin, e.g., **chemotaxis**, **phototaxis**, **thermotaxis** (Chinese **qūyàoxìng**, **qūguāngxìng**, **qūrèxìng**). 趋药性、趋光性、趋热性.

There are three consequences of this difference. (1) The specific taxonomic organization is more explicit in Chinese than in English. (2) The general principle of taxonomy is therefore more accessible (that's the way the Chinese world is organized). (3) There is less *distance* between the world of science and the everyday world than there is in English. Both languages construct basic scientific taxonomies; but in English the general principle is less explicit, the particular structure is less explicit, and (this aspect of) scientific knowledge is further removed from experience of common sense.

When it comes to more elaborated taxonomies, both languages operate on the same principle: the expansion of the noun into a nominal group by premodification. So in both English and Chinese we find the same patterns:

plastic	sùliào	塑料
foam plastic	pàomò sùliào	泡沫塑料
thermal plastic	rèsù sùliào	热塑塑料
engineering plastic	gōngchéng sùliào	工程塑料
phenol plastic	fēnquán sùliào	酚醛塑料

The semantic resource is the same in both languages, and both tolerate about the same degree of expansion without introducing further structural mechanisms.

Objectification

The prototypical form of a scientific taxonomy is a hierarchy of classes of objects, like the taxonomies of the plant and animal kingdoms. But many of the entities named by nouns in scientific discourse are not, in fact, objects, but processes or properties like **fission** and **frequency** mentioned above. In both Chinese and English the noun is a clearly defined grammatical category; in both languages its core meaning is that of an object — an object possessing certain properties and capable of participating in certain processes. Thus in

diamond is energetically unstable/transforms into graphite

diamond is an object; **diamond** [has the property:] **is energetically unstable; diamond** [participates in the process:] **transforms into graphite**. Here **unstable**, representing the property, is an adjective; **transforms**, representing the process, is a verb.

Both these may, however, be represented as nouns: **the energetic instability of diamond, the transformation of diamond into graphite. Instability, transformation** are now 'objectified': that is, they are presented by the grammar as if they were classes of objects. And typically, just as in the history of the *language* these things were first adjectives or verbs and then became nouns (the process may have taken place at any time from early Greek science to the present day), so in the history of a *text* we usually meet them in the same order, e.g.,:

... (diamond is) energetically unstable ... (its) energetic instability (diamond) transforms into graphite ... (its) transformation into graphite

These noun-forming resources (derivational morphology) developed in ancient Greek, were taken over and extended in Latin, then taken over and further extended in English and other European vernaculars.

Chinese also objectifies. So from **bùwěn** 不稳 'unstable' we have **bùwěnxìng** 不稳性 'instability'. In Chinese, however, when a word changes its grammatical class in the syntax, it is not usually signalled morphologically; so

biànhuà 变化 'transform' is also 'transformation'. Thus while Chinese speakers often regard the large number of nominalizations ending in **xìng** 性 as foreignisms, it is not the nominalization itself they are reacting to but its formal marking — the way the noun **xìng** 性 'quality' has become delexicalized through excessive use. It is often said that Chinese does not go in for nominalizations the way that English does; it prefers verbs. It may very well be that in other registers this process of nominalization has not gone so far as in English; but as far as the discourse of science is concerned there is no noticeable difference. For example:

> liǎng diànzǐ-di jìngdiàn chìlì shìnéng wéi zhèng zhí, qiě yǔ liǎngzhě-di
> jiānjù chéng fǎnbǐ
> 两电子的静电斥力势能为正值，且与两者的间距成反比

> the electrostatic repulsion potential of two electrons is a positive value,
> and is in inverse proportion to the distance between them

We refer to this process of objectifying as 'grammatical metaphor'. A semantic feature that is typically realized by one grammatical means comes instead to be realized by another. Thus processes (events, actions, states) are typically realized, in both languages, by verbs; but by grammatical metaphor they come to be realized as nouns. There are very many kinds of grammatical metaphor in both English and Chinese; but the one that dominates the language of science is nominalization. Phenomena which are not objects become 'objectified' by being realized as nouns. Since the core meaning of the category of noun is a class of object, when anything is realized as a noun it takes on some of this object-like status.

Complex Nominalization

In an expression such as **energetic instability** the above features are combined: (1) **unstable** has become an 'object' **instability**; (2) this noun functions as 'Thing' in a nominal group, expanded by a premodifying word **energetic**; (3) **energetic instability** contrasts with **kinetic instability** in a taxonomic relation — two kinds of **instability**; (4) and **energetic** itself embodies grammatical metaphor: grammatically it is not Epithet, as in **an energetic person**, but Classifier '(unstable) with respect to energy' vs. **kinetic** '(unstable) with respect to movement'. I do not know these terms in Chinese, but I imagine that I could coin **néngliàng bùwěnxìng** 能量不稳性 for **energetic instability** and **yùndòng bùwěnxìng** 运动不稳性 for **kinetic instability**, using exactly the same grammatical resource — grammatical energy, or **yǔfǎ néngliàng**, 语法能量, one might say.

The next step is to nominalize not only the process but also any participants and circumstances that go with it: grammatically, expanding the nominal group still further to include some or all of the elements of a clause; for example:

> diànzǐ de guǐdào yùndòng
> 电子的轨道运动

> the orbital motion of an electron

(see the clause **an electron moves in an orbit, diànzǐ shùn guǐdào ér dòng** 电子顺轨道而动). Here by a complex grammatical metaphor the process 'move' has become a noun (Thing), the actor in that Process 'electron' has become its possessor (Qualifier), and the circumstance 'along an orbit' has become the property of the noun (Classifier). Such nominals can of course become very long; the actual Chinese example was:

> diànzǐ-di zìxuán yùndòng hé guǐdào yùndòng xiānghù ǒuhé chéngzǒng-di yùndòng
> 电子的自旋运动和轨道运动相互耦合成总的运动

> 'the combined motion of an electron resulting from the coincidence of the orbital with the rotational motion'

At this point, English and Chinese begin once again to diverge. (Remember they were different in step 1, the taxonomic structure of the lexicon, but similar in step 2, the expansion of the nominal group, and grammatical metaphor.) This time, however, the relationship is reversed: Chinese is *less* explicit than English. Let me explain this. When a process is represented as it typically is — in the form of a clause, the semantic relationships among the components are made explicit by the grammar; e.g., **an electron moves around a nucleus in a circular orbit**. As the process comes to be nominalized these semantic relationships become less explicit: **orbital motion** is a kind of motion having something to do with an orbit, but it is not made explicit whether it is motion of an orbit, in an orbit, caused by an orbit or what. In **movement of an electron**, we are told that the electron is a participant, but not whether it is Actor or Goal (does it move or is it moved?). Now, in English, a typical nominalization of this kind includes both premodifiers, which leave the relationship entirely implicit (see **electrical energy, electrical coil, electrical calibration, electrical potential**, where the semantic function of **electrical** is different in each case), and postmodifiers with prepositional phrases, where the preposition gives some semantic information. A typical example would be **the overall enthalpy charge for the conversion of graphite to carbon dioxide**.

In Chinese, on the other hand, there is no postmodification; all modifiers precede the Head noun. This does not mean that all the transitivity relations of the process are lost in nominalization, because both embedded clauses and (the Chinese equivalent of) prepositional phrases occur in this premodifying position, and such clauses and phrases retain their transitivity structure. Here is a text example:

> yīnér, duìliúcéng dàqì àn qí wùlǐ xìngzhì zài shuǐpíng fāngxiàng-shàng-di chàyì fēnwéi ruògān qìtuán
> 因而，对流层大气按其物理性质在水平方向上的差异分为若干气团

> for that reason the tropospheric atmosphere is divided into several air masses according to differences in its physical properties on the horizontal plane

(note **àn** 按 'according to', **zài. . .shàng** 在…上 'on'). Most of the time, however, these relationships are left implicit, and the only structure markers introduced into the premodifier are **di** 的 showing hypotaxis, and **hé** 和 showing parataxis, equivalent to generalized 'of' and 'and'. Here is an example:

hànzì   bùjiàn   jiégòu   cānshù   fēnxì   di   nèiróng,
character   component   structure   parameter   analysis   'of'   content
fāngfǎ   hé   yuánzé
method   'and'   principle
汉字部件结构参数分析的内容，方法和原则

content, method and principle of the parametric analysis of the composition and structure of [Chinese] characters

In fact there are numerous other possible English equivalents of this nominal group, some of which do not alter the semantic relationships but others of which do, e.g., **analysis of the componential and structural parameters of Chinese characters**. Where English leaves out all **ands** except the last (**a, b, c and d**), but inserts all **ofs**, Chinese leaves out the pre-final **ofs** as well; hence any sequence **a b c . . . di n** permits 2^n-1 different bracketings; and while most of them may not make sense, there is often a subset which do, and a student (whether a foreign student of Chinese, or a Chinese student of science!) may be hard put to it to know which to reject. Furthermore, while English can distribute the elements of a nominal group into pre- and post-modifier ([**a cow is**] **a cud-chewing quadruped which eats grass, a grass-eating ruminant with four legs, a four-legged herbivore which ruminates**, etc. etc.), in Chinese all must precede, so that even the fact that something is a modifier may not be made explicit until right at the end. The sentence following the 'troposphere' example (see p. 129) above is:

[qìtuán ziùshì] zài shuǐpíng fànwéi kě dá shùqiān gōnglǐ ér wēndù, shīdù děng wùlǐ xìngzhì zài shuǐpíng fāngxiàng-shàng-di chàyì bú dà di yí dà kuài kōngqìcéng
气团就是在水平范围可达数千公里而温度，湿度等物理性质在水平方向上的差异不大的一大块空气层

[an air mass is] a large atmospheric layer in which such physical properties as temperature and humidity display little variation in a horizontal direction over a horizontal range of up to several thousand kilometres

— where everything outside the square bracket is a single nominal group with a Head noun **kōngqìcéng** 空气层 'layer of air' at the end. Let me make it quite clear that there is of course no virtue in putting the modifying elements at one side of the Head noun or the other. What is relevant is that, whatever the resources in a particular language for creating indefinitely expandable nominalizations of this type, they will be put under maximum pressure by the requirements of scientific discourse. Each language will then respond in its own way; and in Chinese the effect is to create a high level of ambiguity and to leave implicit many of the semantic relationships among the various elements of the process.

The Functions of Nominalization

Our last point, then, will be to ask: why does this pressure arise? Why does scientific discourse demand this very high degree of nominalization with all the grammatical metaphor that this involves?

There seem to be two main reasons. One concerns the structure of scientific argument, as developed in Europe by Isaac Newton and his successors. The core of a scientific text was the development of a chain of reasoning (ultimately based on experiments) in which each step led on to the next. But in order to lead on to the next step you have to be able to repeat what has gone before and is now being used as the springboard for the next move. Notice how the grammar does this, with an English and a Chinese example:

. . .both ethyne and nitrogen oxide are kinetically stable . . . . .
The kinetic stability of nitrogen oxide shows . . . .

tiānqì shì shùnxíwànbiàn-di; dàn tādi biànhuà shì yǒu ídìng guīlǜ-di
天气是瞬息万变的；但他的变化是有一定规率的

the weather is constantly changing; but its changes have a definite pattern

In both languages the grammar 'packages' what has gone before by nominalizing the Process (attribute or event), and making the Medium of that Process a 'possessive' modifier. This enables it to function as the Theme of the succeeding clause. This is the simple, most basic form. Sometimes however the new step in the argument has to include not just the previous one but a great deal that has gone before (and sometimes to anticipate much of what is to follow); and this may require a more complex package, picking up a number of related motifs; e.g.,

The great reactivity of fluorine in these reactions with non-metals [is explained . . .]

pǔtōng yíngguāngguǎn suǒ fāshè-di guāng shì yǒu zhèzhǒng zǐwài yuèqiān suǒ jīfā-di túliào-di yíngguāng
普通荧光管所发射的光是有这种紫外跃迁所激发的涂料的荧光

the light that is emitted by the usual fluorescent tube is a fluorescence of the coating stimulated by this kind of ultraviolet transition

The noun is the only syntactic class that can accept thematic 'packages' of this kind; or, to put this another way round, the process of packaging turns (metaphorizes) all events and attributes into textual objects. I have discussed this phenomenon in another paper (see Chapter 3) and will not try to elaborate it here. The second reason for nominalizing has to do with the structure of scientific *knowledge*. While the argument has to be dynamic (hence the flow of the text), the edifice that is constructed by it is a static one — or let us say that it embodies a synoptic rather than a dynamic representation of reality. Newtonian science has to hold the world still, to anaesthetize it so to speak, while dissecting it — if you are trying to understand something, then in the early stages of your enquiry it is helpful if it does not change while you are examining it. Of course, reality does change while you are looking at it, and twentieth century science is coming to terms with this; so the grammar of scientific discourse in the next 500 years will probably be very different from what it is now — perhaps more like that of spoken language with its clauses rather than expanded nouns. To begin with, however, scientists had to create a universe that was made of *things*. The concept of a taxonomy, that we started with, is first of all a taxonomy of objects; it is only later that other kinds of phenomena come to be classified in this way, and again in order for this to happen the grammar has to turn them into nouns (they are types of **change**, **motion**, etc.; see Chapter 8).

So the grammar of both Chinese and English, using similar resources (noun compounding, the nominal group, nominalization and grammatical metaphor), creates a form of discourse for codifying, extending and transmitting scientific knowledge. With this discourse, the *argument* — the rhetorical movement — is made very clear and explicit; while the *content* — the conceptual structure and internal relationships — may be left highly implicit, in one way or another (each language in its own way). We are not saying that for the meaning to be implicit in this way is something undesirable; but this kind of grammar does make demands on both writer and reader: demands on the writer to ensure that the text provides the semantic information that the reader needs in order to construct the taxonomies, decode the metaphors, and follow the argument; and demands on the reader to be alert enough to receive and make use of this information.

What seems important to me, as a linguist and as a teacher, is that we should understand both *how* the grammar of science works and *why* it works the way it does. For this we need to examine scientific discourse, analysing the grammar in terms of its different functions ('metafunctions', in the systemic sense; especially the ideational and the textual); and also to trace its history, how it evolved the way it did. I asked at the beginning: would scientific Chinese have been different if modern science had evolved first in China? Obviously we can never answer this kind of question with certainty. But technology *did* evolve in China, and was ahead of technology in Europe throughout most of history; so we can compare the earlier forms of scientific discourse that developed in classical Chinese and in medieval Latin (and also the early texts in English and other European vernaculars). Of course, although we can ask why things happened in the past, we cannot change them. We can, however, adopt practices that will have an effect on the future; and, as linguists, and discourse specialists, we have a responsibility to look ahead: to assist those who are learning science either in their mother tongue or through some foreign language; to advise on language planning — whether scientific language should be 'engineered' in some way (as is done with chemical terminology, for example; but not limited to the creation of technical terms); and, not least, to raise issues of the role of science, and of scientists, in our modern society. If we examine the discourse of science we can become more aware of the ideology that is enshrined in the way scientific language construes the world.

School Literacy: Construing Knowledge

Introduction

Part 2 of this volume reports on research into the secondary-school discourses of science and humanities conducted in and around the Department of Linguistics at the University of Sydney from 1986 to 1990. Throughout this period Joan Rothery provided ongoing support as far as the educational implications of the work were concerned; in 1986 and 1987 funding provided by the Australian Research Council made it possible for Suzanne Eggins, Radan Martinec and Peter Wignell to work alongside Martin on the research. The joint authorship of two of the chapters reflects the dialogism of this enterprise; and all of the papers in Part 2 draw heavily on the insights of this research team. The work is part of an ongoing longer term research program addressing literacy issues in Australian schools initiated by Martin and Rothery in 1979; very readable introductions to practical aspects of this work are found in Derewianka 1989 and Littlefair 1991 (for a brief overview of the work in general see Martin, 1991).

Chapters 8 and 9 present the results of research into the discourse of geography and science as these are taught in Australian secondary schools. The first chapter, 'The discourse of Geography . . .', concentrates on physical geography and was written with an audience of educational linguists in mind. It is particularly concerned to deconstruct physical geography from the perspective of field, paying careful attention to the language used to construct this kind of un-commonsense perspective on our world. As far as we could see at the time, in broad terms this un-commonsense was organized in two complementary ways:

1 it was comprised of un-commonsense *taxonomies* of physical entities and to a more limited extent of processes; these taxonomies organized entities into class/sub-class (superordination) and into part/sub-part (composition) relations.
2 at the same time it was comprised of un-commonsense *implication sequences* which provided step by step explanations of physical processes.

In order to establish these un-commonsense taxonomies and implication sequences, we noted that geography had to translate commonsense perspectives into un-commonsense ones; to do this, it made use of technical terms. The paper explores in some detail the ways in which these technical terms were built up by the grammar and the important function they have in consolidating information, which we referred to as *distillation* (similar to thematic condensation in Lemke, 1990a; pp. 95–6).

The second chapter, 'Literacy in Science . . .', concentrates on secondary-science discourse and was written for an audience of science educators. This work draws on important work by Natalie Shea in 1988 into the language of Australian junior secondary textbooks. It confirms that the model of taxonomies, implication sequences and technicality developed for physical geography can be generalized across the physical and biological sciences. In addition it addresses science discourse from the perspective of genre as well as field, noting the two main genres deployed to construct un-commonsense knowledge — *report* (used to build up taxonomies) and *explanation* (used to build up implication sequences). A number of other genres used in secondary science, including experiment, exposition and narrative are also briefly considered.

This chapter is also concerned in part with the pedagogical implications of the research. Like Lemke, 1990a; pp. 169–174, the importance of *writing science* in order to consolidate un-commonsense knowledge, alongside speaking science, is emphasized. This is especially important in the Australian context, as elsewhere, because students do so little science writing in schools; the problem is compounded in some classrooms where science textbooks are no longer used, with the result that suitable models of written science discourse are no longer readily available. When we take into account that fact that the un-commonsense knowledge of science has evolved over the centuries in written text, the tremendous political significance of this functional illiteracy rears its ugly head. The chapter is also critical of progressive trends in science pedagogy which place undue emphasis on genres such as narrative which have not evolved to construct un-commonsense; these trends run the risk of stranding students in commonsense discourses even after several years of study in science classrooms. Readers interested in following up this critique within the context of Australian literacy debates will find the following references helpful: Christie *et al.*, 1989; Cranny-Francis *et al.*, 1991; Giblett and O'Carroll, 1990; Gray, 1987; Hammond, 1987; Lee and Green, 1990; Martin, 1991; Martin *et al.*, 1987; Painter, 1986; Painter and Martin, 1986; Reid, 1987; Rothery, 1989a, 1989b; Thibault, 1989; and Threadgold, 1988, 1989.

Chapters 10 and 11 recontextualize the work on physical geography and science introduced above with respect to the structure of humanities discourse in secondary school, in particular the discourse of history. The first chapter, 'Technicality and Abstraction . . .' was written for student teachers and introduces the basic distinction between *technical* (i.e., scientific) and *abstract* (i.e., humanities) writing. It makes the important point that both of these discourses rely heavily on the metaphorical grammar of written English, although they deploy this resource in different ways. The second chapter, 'Life as a Noun . . .'*, takes the discussion of technicality and abstraction a step further, with an audience of professional discourse analysts in mind. This is particularly concerned to outline the ways in which abstract humanities writing uses grammatical metaphor to construct what Halliday has referred to as 'hierarchy of periodicity' in text (in Thibault, 1987, p. 612) — waves of Theme/Rheme and Given/New structure organizing ascending levels of compositional structure. For further discussion of this 'text like a clause' motif, see Halliday 1981, 1982 and Matthiessen, 1988.

* This is from a paper first delivered as a plenary address to the 16th International Systemic Congress held in Helsinki, Finland in 1989.

Chapter 8

The Discourse of Geography: Ordering and Explaining the Experiential World*

P. Wignell, J.R. Martin and S. Eggins

Geography in the Curriculum

In New South Wales secondary schools, physics, chemistry, biology, and chemistry have been taught as an integrated subject, Science, since the mid-sixties. Geography remains a separate subject area, with a focus on physical geography in the junior secondary school. For the most part its discourse is indistinguishable from that used in science (Martin, 1990b), the main difference being the absence of experiments to illustrate the scientific world view constructed in introductory textbooks. The most closely related discipline in American high schools would appear to be Earth Science.

Geography as explained to junior high-school students is very explicit about its goals. One junior high-school textbook (Sale, Wilson, and Friedman, 1980, p. 3) states that the primary task of the geographer is to look for 'order and meaning in the world'. The procedures used to uncover this order and meaning are, first, to 'observe and describe'; then, 'to group and classify' and finally 'to analyse and explain'. 'Analyse' is further defined as 'to seek an explanation'. The three stages of the geographer's task can thus be summarized as observing, ordering, and explaining the experiential world.

The New South Wales School Certificate Syllabus in Geography (Secondary Schools Board, 1984–1985, p. 6) lists the 'important questions' geographers ask. These are:

What is there?
Where is it?
Why is it there?
How is it changing over time?
What are the effects of its being there?

If we try to fit the 'grouping and classifying', and 'analysing and explaining' of the textbook to the above questions, we find that, broadly speaking, the first two

* This chapter is taken from: *Linguistics and Education* (1989), **1**,4, New Jersey, Ablex pp. 359–91.

questions fit the grouping and classifying task. That is, things can be classified according to where they are or what they are: For example, one can have tropical climates, mid-latitude climates, and polar climates; one can also have mesas and buttes. The remaining three questions fit the analysing and explaining task. That is, things can be explained according to what caused them, and how they effect other things.

Through our study of junior high-school geography textbooks, we have tried to develop a description of the discourse of geography: That is, how language is used to represent and teach the field of geography, and the geographer's task.

Our analysis here suggests that in the discourse of geography, language is used in three distinctive ways, which corresponds to the three tasks geography sees itself as fulfilling. First, language is used to 'observe' the experiential world through the creation of a technical vocabulary: a process of dividing up and naming those parts of the world which are significant to geographers. Second, language is used to order the experiential world, through the setting up of field-specific taxonomies. And third, language is used to 'explain' the experiential world, through the positing of implicational relations among natural or manmade states.

In the language of geography only the functions of classifying and explaining necessarily involve a shift from the common sense or everyday use of language. Ordinary, everyday language can be and is used to observe and describe things. For example a landform could be described as a kind of a hill with steep sides and a flat top. But by naming that landform a mesa, the geographer has created a technical name for the landform and at the same time placed it into a set of taxonomic oppositions with other kinds of landforms, for example, mesas and buttes.

Naming a thing always implies a classification, and the same thing with the same name can be classified differently depending on who is doing the classifying. For example, to the person in the street a pumpkin would probably be classified as a vegetable: something belonging to that class of things which people eat, which grow on plants, and which are usually cooked first (things not cooked are either fruit, or salad vegetables). To a botanist a pumpkin would be a fruit, because to a botanist fruit means the part of the plant which contains the seeds.

In science the move from describing to classifying is a move from the everyday to the technical; and the move from classifying to explaining is a move from talking about things to talking about processes.

In this article the linguistic resources that are used to achieve these tasks are examined. In so doing, a description of the discourse of geography will be offered, and a model of a technical/scientific field will be developed.

Taxonomies

In order to understand how geography uses language to observe and order the experiential world, it is necessary to begin by considering briefly the nature of the 'taxonomy' in the natural sciences.

A 'taxonomy' is an ordered, systematic classification of some phenomena based on the fundamental principles of superordination (where something is a type of or kind of something else) or composition (where something is a part of something else).

Table 8.1: *Botanical Classification of One Variety of Rose*

	Latin	English
Division	Spermatophyta	Seed plants
Class	Angiospermae	Flowering plants
Subclass	Dicotyledoneae	Dicots
Order	Rosales	Rose order
Family	Rosaceae	Rose family
Genus	Rosa	Rose
Species	Rosa setigera	Wild climbing rose
Variety	Rosa setigera tomentosa	A special wild climbing rose

Source: From Benson (1957) p. 3.

When one thinks of taxonomies, what is called to mind is probably the kind of taxonomy found in the natural sciences, botany for instance. The taxonomies found in botany are based on a system first proposed by Linneaus (1753, 1957) in *Species Plantarum*. Following the Linnaean system of classification (with later modifications) every plant in the world can be put into order, or in its place in an overall system (e.g., see Table 8.1). This system is based on superordination: That is, where something is a kind of or type of something else. So, a wild climbing rose is a kind of rose . . . is a kind of dicot, is a kind of flowering plant, and so forth. One also finds composition taxonomies in botany: that is, where something is classified as being a part of something else. For example, plants can be divided into their component parts (roots, stem, buds, leaves).

Taxonomies like this replaced earlier classification systems which gave a detailed description of the plant. For example, the species described under the Linnaean system as *Physalis angulata* was previously described as:

> *Physalis annua ramosissima, ramis angulosis glabris, foliis dentato-serratis.* (A physalis with branches growing in circles from the trunk, angular, hairless branches and leaves serrated like teeth.) (Linneaus, 1753, 1957, p. 2)

This description tells a lot about the plant, but nothing about its relationship to other plants. The binomial system, by contrast, tells little about the plant (unless one knows the criteria upon which the classification is based), but enables it to be placed in an ordered system of oppositions in relation to all other plants.

In adopting the binomial system, what Linnaeus in fact did, was draw upon the way in which everyday language is used to order things. For, although taxonomizing in the natural sciences is both explicit and formalized, taxonomizing itself is not restricted to the natural sciences or even to science. It is a characteristic of all fields, including everyday or nontechnical ones, that the parts of the world concerning them are ordered into taxonomies.

For example, a vernacular taxonomy of the parts of the human body relevant to bodybuilders is shown in Figure 8.1.

Figure 8.1 is a composition taxonomy, not listed in any textbook, but drawn on by anyone who has done any serious weight training. It would contrast quite markedly in its opposition with the organization of the same body parts taken from an anatomy textbook, for example.

Figure 8.1:   Vernacular Composition Taxonomy of Body Parts for Bodybuilders

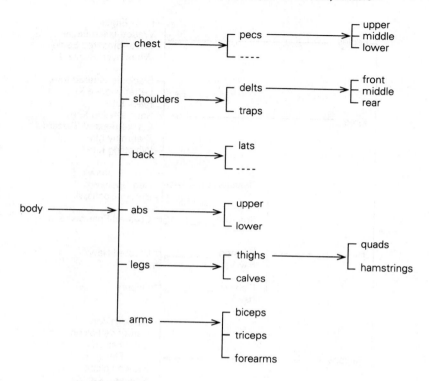

This illustrates both the existence of implicit vernacular taxonomies, and how different fields will name, reorder, or reclassify similar things differently, according to what is 'emic' (meaningful or relevant) to that field. Notice that only major groups of skeletal muscles are taxonomist. Things such as the head, hands, and feet, which would normally be thought of as body parts, are not included because they are not relevant to the field.

Specialized vernacular taxonomies can also be explicit and can exist alongside formal scientific taxonomies of the same field. For example, Slater's (1983) *A Field Guide to Australian Birds*, provides both a scientific taxonomy based on the Linnaean system and a folk taxonomy. The folk taxonomy for birds of prey is given in Figure 8.2.

The folk, or vernacular, taxonomy in Figure 8.2 is really a very informed kind of folk taxonomy used by bird-watchers to identify birds. It is different from both a lay person's taxonomy and from a scientific taxonomy. For example, when a group of people with no assumed knowledge of birds was asked to construct a taxonomy of birds of prey, the result was something like Figure 8.3. These two vernacular taxonomies can be contrasted with a scientific taxonomy of the same birds of prey, shown in Figure 8.4a.

By examining three taxonomies one can see where they are alike and not alike. Note, first of all, that the differences between the folk taxonomy (Figure

Figure 8.2: *Birdwatchers' Vernacular Taxonomy of Birds of Prey (Raptors)*

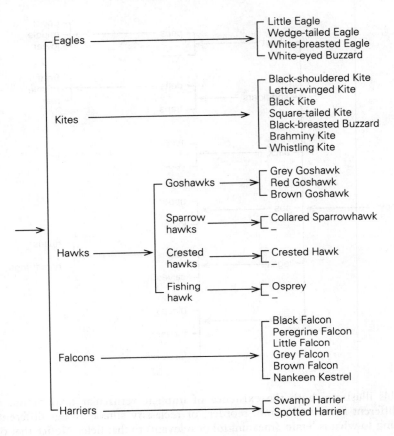

Source: adapted from Slater 1983

8.3) and the bird-watcher's taxonomy (Figure 8.2) are far greater than the differences between the bird-watcher's and the scientific taxonomy (Figure 8.4).

The naive or uninformed taxonomy (Figure 8.3) seems to be based upon the criteria that any bird which catches live animals other than insects is a bird of prey. In fact, when there are no mice around, a major part of a nankeen kestrel's (a small falcon) diet is grasshoppers. This taxonomy is also rather indelicate; it gives no sub-classifications of hawks or eagles.

The bird-watcher's taxonomy (Figure 8.2) is far more delicate. It identifies twenty-five species, divides them into eight genera (five of which are named: kites, goshawks, sparrow hawks, crested hawks, and fishing hawks). It also classifies the species into four families: eagles, hawks, falcons, and harriers.

The criteria for classification, particularly at the level of species, are usually physical characteristics:

Fig 8.3:   *Uninformed Vernacular Taxonomy of Birds of Prey*

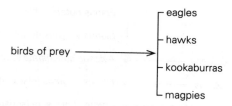

Figure 8.4a: *Scientific Taxonomy of Birds of Prey (Falconiformes)*

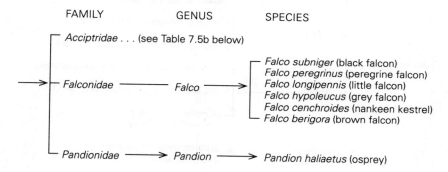

colour	black falcon
markings	spotted harrier, white-breasted sea eagle
size	little eagle
habitat	swamp harrier
sound	whistling kite
peculiarity	crested hawk

The similarity between the two vernacular taxonomies (Figures 8.2 and 8.3) is that classification is based upon criteria which are readily observable.

If the bird-watchers' and the ornithologists' taxonomies (Figures 8.3 and 8.4) are now compared, it is found that, although some of the names are either direct translations into Latin (*Hamirustra melanosterna* = black-breasted buzzard), or close approximations (*Elanus scriptus* = letter-winged kite), the way the birds are ordered differs. The same twenty-five species are now classified into twelve genera and three families. The criteria for this scientific classification are different from those used in the vernacular taxonomies. The birds are classified according to similarities in chromosomes and genes — things not readily observable without sophisticated equipment. No birds have been added to or subtracted from the list, and most of the reordering has been done at the more superordinate levels of the taxonomy.

The three taxonomies above seem to be operating on a cline of needing to know. The person in the street probably knows the names of a few species of

141

Figure 8.4b: Scientific Taxonomy of Birds of Prey (Acciptridae)

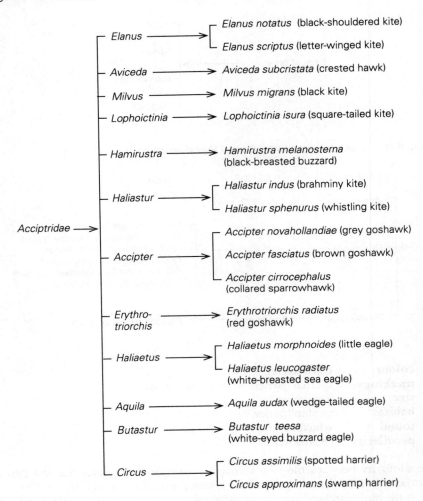

birds, but doesn't need to know anymore. Serious bird-watchers could not operate with the lay person's taxonomy. They need to know more, and in knowing more, also have a different way of knowing. The bird-watcher, the person in the street, and the ornithologist see the same birds, but they see them differently.

One way of classifying the different types of taxonomy reviewed here would be to oppose scientific and vernacular taxonomies in terms of whether technology is used to augment the senses when making the observations that form the basis of the classification. Within the vernacular taxonomies, the more delicate specialized taxonomies could then be distinguished from general everyday ones (see Figure 8.5). This comparison between scientific and vernacular taxonomies brings out the point that phenomena classified formally and scientifically often

Figure 8.5:   Scientific and Vernacular Taxonomies

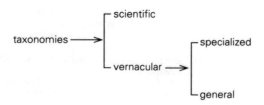

already have vernacular names and vernacular classifications. Much scientific taxonomizing, then, is a process of renaming in order to reclassify the vernacular. That is, a binomial Latin name is substituted for a vernacular name; and then the renamed phenomena are given a scientific ordering which will be distinct from the existing vernacular ordering.

This is not to suggest that a formal or scientific taxonomy is just a renaming of an existing vernacular one. Technical language cannot simply be dismissed as jargon, because alongside a renaming, there is also a reordering of things. The two informed taxonomies of birds of prey name exactly the birds, but there are substantial differences in the way they are ordered.

Geography, like all other fields of human interest, organizes phenomena into taxonomies. The phenomena that concern geographers (the things and processes of the experiential world) very often already have existing vernacular names. For example, the taxonomy of climate shown in Figure 8.6 uses vernacular terms but arranges them in oppositions specific to geography: The terms may be vernacular, but their taxonomic organization is not.

In Figure 8.6, the composition taxonomy acts as a list of ingredients, which in different proportions, realizes the different kinds of climate listed in the superordination taxonomy.

As stated above, scientific taxonomizing typically involves two steps: the renaming of vernacular terms in order to then reclassify them into scientific taxonomies. A precondition of taxonomizing is the naming of things to be ordered. In the natural sciences, the process of naming is highly formalized. However, in fields such as geography, where the Latin binomial system is not used, grammatical resources are used to set up the names for things in the taxonomies.

Setting up this technical vocabulary, to enable the ordering and classification of the experiential world, is a major part of the discourse of geography, which will now be considered under the heading of technicality.

Classifying the Experiential World: Technicality

The possibility of ordering the things of the experiential world in some field-specific way presupposes both observing and naming relevant phenomena. Observation may be, in part, an empirical and nonlinguistic activity, but the record of observation is always a linguistic one: It involves giving things names.

Figure 8.6: *Geographical Taxonomy of Climate*

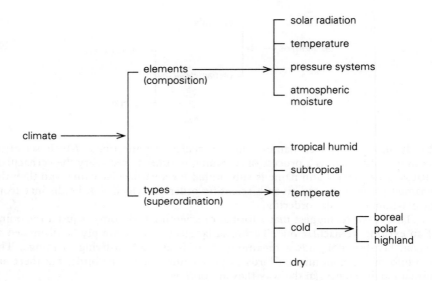

Obviously, many of the phenomena the geographer observes and names, such as lakes and deserts and rainforests, have already been observed and named by other people. The nongeographer can attach the name desert to the appropriate natural phenomenon without much difficulty; he or she probably thinks of the name desert as referring to some vast, dry area covered with sand. But for the geographer, the term desert refers to 'an almost barren tract of land in which the precipitation is so scanty or spasmodic that it will not adequately support vegetation. The salient feature of this type of climate is aridity, or lack of regular precipitation'. (Moore, 1949, 1966, pp. 53–4). When used by a geographer, then, the term desert has a field-specific meaning. When names are given field-specific meanings in this way they become what are called technical terms.

In geography most technical terms are nouns. There are some technical verbs, such as condense, but if they are used they usually occur in close proximity to their nominalized form (e.g., condensation).

For example, (unless otherwise stated all examples are taken from Sale *et al.*, 1980, and the technical terms appear in boldface):

> 1. When the temperature of saturated air falls as it does when it rises in a convection current the water vapour turns into tiny water droplets or, as we say, it **condenses**. Further **condensation** . . . . .

Here the technical term is introduced as a verb (process) and then immediately treated as a noun (thing).

Technicality, as the term is used here, refers to the use of terms or expressions (but mostly nominal group constituents) with a specialized field-specific meaning. For example, the term duck has a different meaning for the cricketer (e.g., out for a duck) as opposed to the bridge player (e.g., to duck a trick), or the haberdasher (e.g., a kind of cloth); and none of these meanings will be equivalent to the commonsense vernacular meaning (a bird with webbed feet and a flat beak).

When we say that a term won't have the same meaning, what we're really saying is that it won't enter into the same taxonomic oppositions. For example, the term grammar probably enters into a vernacular taxonomy of speaking, in opposition to accent, whereas within the field of linguistics, where grammar is a technical term, it enters into oppositions with phonology, lexis, semantics, and so on.

The process of technicalizing — for example, of building up a technical vocabulary — involves two steps: (a) naming the phenomenon, and (b) making that name technical. Each of these steps will be considered in turn.

Naming Phenomena

The first step in the process is to find a name for the phenomena. Terms being set up as technical within the field of geography can overlap with terms from other fields in a number of ways:

a. Some technical terms are words which already have a vernacular meaning, for example, environment, rain, wind. The purpose of the text is to reassign a new *valeur*, or meaning, to these terms (to fit them into a geographer's taxonomy of weather, rather than a vernacular one).

b   Some technical terms are words which are also used technically in other (related) fields, for example, condensation, transpiration, protozoa.

c. Some technical terms are words which are indexical of that field (i.e., unlikely to have a vernacular or other-field usage), for example, mesas, buttes, wadis.

Whether these words are borrowed from the vernacular language or the language of another field, when they are set up as technical terms within geography they acquire a meaning, or *valeur*, specific to that field (for a discussion of *valeur*, see Saussure, 1915, 1966).

Most technical terms are nominal group constituents, usually things or Classifier^Thing compounds (for discussion of the nominal group, see Halliday, 1985a, pp. 159–74). This is obviously not accidental, for in order to classify and organize with language, we need first of all to turn phenomena into things or nouns. The grammar has extensive resources within the clause and nominal group structure for organizing things, but very limited resources for organizing processes.

This need to turn phenomena into things or nominals means technical terms may be derived in a number of ways.

a. A technical term may be a single nominal or thing, for example, mesas, buttes, herbivores, consumers. These types of technical terms tend to be the names for physical or living objects, and also tend to be things for which vernacular taxonomies exist, although, less delicate ones.

b.   A technical term may consist of a nominal group compound, with a Classifier^Thing structure. These compounds tend to be ones that are familiar, but which, in the vernacular, appear descriptive rather than classificatory (Epithet^Thing structures in Halliday's terms), for example, physical environment, raw materials, balanced state, water vapour.

c.   Technical nominal group compounds of the Classifier^Thing type may also be derived from what we call implication sequences. These sequences, which will be discussed in detail later, are the way geography uses the resources of the clause complex to posit cause and effect between phenomena, for example, relief rainfall, frontal rainfall. In these compounds, where only the thing may have a vernacular usage, the Classifier represents the agent in an implication sequence (i.e., that which caused the rain to fall).

It can be seen that there's a kind of progression here, away from discrete physical objects, towards more complex phenomena being turned into things.

d.   A technical term may also be derived through nominalization; that is, turning happenings into things which can be technicalized. This is where grammatical metaphor, specifically nominalization, gets mobilized (for discussion of grammatical metaphor, see Eggins, Wignell, and Martin, 1987; Halliday, 1985a). Thus, some technical terms are single nominals or things but realize a nominalization, for example, condensation, transpiration.

Here the process and medium of the implication sequence have been conflated into a nominalization of the process. Rather than just naming things, the processes by which things come to be are now being named.

e.   An extension of this is to have a technical nominal group compound with a Classifier^Thing structure, but where the Classifier is a nominalization representing the agent from an implication sequence, for example, convection currents.

f.   Finally, processes used as technical terms are found in geography texts, for example, condenses, evaporates.

But technical verbs always have a corresponding nominalized form so that they can be treated as things in the text. For example, the technical process 'condense' will occur alongside, and will imply, the existence of its technical nominal, 'condensation'. These technical verbs are rare, whereas technical nouns proliferate.

Making Names Technical

If the first step in the technicalizing process is to turn phenomena into named things, the second is to set up these names as technical. An important means of doing this is by signalling or marking terms which are going to be given a technical status.

Table 8.2: Taxonomic Organization of a Geography Textbook

Book title — Climate

Part title — Part 1 — Elements of Climate
           Part 2 — The World Pattern of Climate
                     Climatic types and their distribution

Chapter title — *Part One*
                Solar Radiation
                Temperature of the Atmosphere
                Wind and Pressure Systems
                Atmospheric Moisture and Precipitation
                Air Masses and Fronts

                Part Two
                Climatic Types and Distribution
                Tropical Humid
                Subtropical
                Temperate
                Boreal, Highland, Polar
                Dry

Source: From Trewartha (1986) table of contents.

In the geography textbooks examined for this study, most technical terms are explicitly marked orthographically in the text the first, and sometimes the second time, they are used. Orthographic marking involves printing the word in italics, capitals, boldface, parentheses, or commas, and so on.

For example, in Text 1 notice the following words printed in boldface: orographic, relief, convectional rainfall, frontal rain.

TEXT 1

Types of rainfall

a. **Orographic** or **relief** rainfall. Moist air from the sea is forced to rise by the mountains. Cooling takes place, clouds form and rain falls on the windward slopes of the mountains. The lee slopes receive little rain because most of the water vapour originally in the air has fallen on the windward slopes.

b. **Convectional rainfall**. Air containing water vapour becomes heated, expands and then rises because it is less dense than the surrounding air. When it reaches a sufficient height it becomes cold again and condensation takes place and rain may fall.

c. **Frontal rain**. In the diagram a moving mass of cold air, being heavy, has pushed under some warner, lighter air. The warm air cools as it is pushed up and clouds and rainfall result (Sale *et al.*, 1980, p. 11).

The tendency in geography textbooks is for terms in capitals to occur as either chapter headings or subsection headings within a chapter, and those printed in boldface type to occur within the body of the text. This seems to correspond to superordinateness in the taxonomy.

For example, the taxonomic organization of the title page of Trewartha's (1968) work on climate is shown in Table 8.2. This could be drawn up into the superordination and composition taxonomies shown in Figure 8.6.

One function of marking technical terms in the texts is to draw the students' attention to important words that will recur, and constitute part of the students' assumed cumulative knowledge. But the principal function of marking terms in this way is to signal that they are about to be given a field-specific meaning. A marked term must therefore be accompanied by a definition, or (to give it its technical name) an elaboration, somewhere in the nearby text. Typically, once the term has been elaborated, it is no longer highlighted. It just becomes another name for another thing.

For example, in Text 2 the terms water vapour, humidity, and relative humidity, are marked the first time they are used, but not the second, because by then they have been elaborated or defined.

TEXT 2

Water in the atmosphere
When water evaporates, it changes into an invisible gas called **water vapour**. When air contains a lot of water vapour, we say that the **humidity** is high. At different temperatures, air can hold different amounts of water vapour. When the air cannot hold any more water vapour at a particular temperature, we say that the air is **saturated** or that it has a **relative humidity** of 100 per cent. If the air is carrying only three quarters of the water it is capable of carrying, then it has a relative humidity of 75 per cent; if it is carrying half its capacity, then its relative humidity is 50 per cent, and so on. Relative humidity, then, is a measure of the amount of water vapour being carried in air of a particular temperature compared with the total it could carry at that temperature.

At high temperatures, air can carry a great amount of water vapour. When the temperature of saturated air falls, as it does when it rises in a convection current, the water vapour turns into tiny droplets of water or, as we say, it **condenses**. Further condensation may lead to the formation of clouds and precipitation in the form of rain, hail or snow. (Sale *et al.*, 1980, p. 11)

Marking a term orthographically in the text is only the first step is the process of setting up technical terms. The second step is to assign to each term its field-specific meaning: the meaning that it will encode whenever it is used again within the context of that field. In fact, once a term has been given technical status through marking, the reader looks for and expects to find some kind of explanation in the text of what the term means or refers to.

Naming and Defining

When looking more closely at the way technical terms are set up in geography textbooks, it is found that this is typically done in two ways: through naming and through defining.

The three grammatical resources of projection, elaboration, and enhancement each provide a different means of assigning a field-specific (in this case geographical) meaning to each new technical term introduced. Through projecting and nonprojecting naming processes, a technical name can be attached to a phenomenon which usually has an existing vernacular name; through elaboration a technical term can be defined according to what it is or stands for; and through enhancement a technical term can be defined according to what caused it to be or happen.

Naming

Technical terms can be named rather than defined using either projecting or nonprojecting naming processes (for projection, see Halliday, 1985a). For example, a number of terms are introduced via a projecting verbal process, such as 'we say':

> 2.a. When air contains a lot of water vapour, we say that the humidity is high.
> b. When the air cannot hold any more water vapour at a particular temperature, we say that the air is saturated or that it has a relative humidity of 100 per cent.

Often terms are introduced through the use of nonprojecting naming processes: is called, is known as, we use the term *x* for.

> 3.a. The common name for a saprophyte is fungus.
> b. Thus green plants are called producers.
> c. Vegetation adapted to withstand drought is known as xerophytic vegetation.

These naming processes make quite explicit that the term is being set up as technical. They also bring out the point that what the process of technicalizing is really about is translating: giving a field-specific gloss to phenomena which may be known as something else in another field or in folk taxonomies.

Defining

The grammar embodies a range of ways of defining or elaborating terms. The most familiar and probably the most frequently used is to define technical terms through an identifying relational clause (Halliday, 1985a). The token is enclosed in braces and the value in square brackets in examples 4 to 12, to follow.

> 4.a. {An ecosystem} is [that home or place in which a community or group of interacting plants and animals lives].
> 4.b. {The biome} is [the living part of the ecosystem].
> 4.c. {The physical environment} provides [the home for the living residents].
> 4.d. {Precipitation} refers to [all forms of water which fall (precipitate) from the sky].
> 4.e. {Prevailing winds} are [winds which blow from one direction for long periods during the year].

In these identifying clauses the meaning is: *x* is defined by *y*, or, *y* serves to define the identity of *x*.

Halliday (1985a, p. 115) points out that the thing that is doing the identifying can specify the identity of the target thing in one of two ways: (a) by specifying its form, how it's recognized; (b) by specifying its function, how it's valued. These two sides to an identifying relationship give the two grammatical functions of token and value. Halliday glosses the value function as realizing the 'meaning, referent, function, status, role', and the token function as realizing the 'sign, name, form, holder, occupant'.

In all the geography and other science texts examined so far, it turns out that the technical term always realizes the function of token, and that what is thought of as the definition realizes the role of value. Tokens can be grammatically distinguished from values in that they are subject in active identifying clauses and complement in passive. For example:

Active: Precipitation refers to all forms of water which fall from the sky.
Passive: All forms of water which fall from the sky are referred to as precipitation.

Although relational processes can be glossed as processes of being, there is a set of processes other than *to be* which can be used to identify, most of which have passives. These include equals, adds up to, makes, comes out at, signifies, means, defines, spells, indicates, expresses, suggests, acts as, symbolizes, plays (a role), represents, stands for, refers to, exemplifies.

The fact that the technical term comes out as token makes technicality look even more like a translating process, since the agnation can be demonstrated among sentences like:

5.a. In French, {*chien*} means [dog].
  b. In French, {*chien*} stands for [dog]
  c. In English, {C–A–T} spells [CAT]
  d. In English, {CAT} means [a feline]
  e. In geography, {the biome} is [the living part of the eco-system].

The token/value relationship seen in these identifying clauses is in fact just one grammatical means of realizing the general relationship of $y = x$ or y defines x. This relationship of elaboration can be realized in a variety of other grammatical ways, to which the labels of Token and Value can also be generalized. Some examples of these other ways of elaborating technical terms include embedded clauses (defining relative clauses):

6.a. When water evaporates it changes into [an invisible gas] called {water vapour}.
6.b. Desert streams usually drain down into [the lowest portions of nearby desert basins] which are called {bolsons}.

elaborating nominal groups:

7.b. at [the lowest level], {(trophic level 1)} at [the next lowest level], {(trophic level 2)} at [the final level], {(trophic level 3)}
7.b. {certain protozoa} [(single-celled animals)]
7.c. . . . {the diurnal range in temperature} [(the daily range between the maximum and minimum temperatures in 24-hour periods)] is great.

and, elaborating conjunctions (group/clause):

8.a. {trophic level 1} [(that is, where life forms are the simplest)]

8.b. [When the air cannot hold any more water vapour at a particular temperature], we say that the air is {saturated} or that it has {a relative humidity of 100 per cent}.

8.c. [Another visible example of desert air motion] is {the small whirl-wind}, {dust-devil}, or {willie-willie}.

8.d. {Orographic} or {relief rainfall}: [Moist air from the . .]

In all the above examples, the token and value relationship is established within one clause. However, one also finds technical terms introduced in one clause or group, but elaborated in one or a number of following clauses. The relationship between the technical term and its subsequent definition is established through reference. For example,

9.a. . . . until eventually all the species of plants and animals in the eco-system live in {a balanced state}
this means [that they depend upon one another and live in harmony unless disturbed].

In this example, the token is realized by the anaphoric referent 'this' (referring back to the technical term 'balanced state'). The value is realized by an embedded clause complex.

Reference can also be used to establish a relationship between a technical term and the activity sequence which produced it:

9.b. You have probably learned the meaning of the term {transpiration} in your science lessons.
[In this process, plants lose water in the form of vapour through their leaves, this water is replaced with water containing plant food collected by the plant roots . . .]

Here, the anaphoric referent 'this' stands for the technical term 'transpiration'. However, instead of defining what the term 'transpiration' means, an explanation of how transpiration occurs is found. The implicit token/value relationship can be brought out when this is reworded as:

Transpiration is when. . . .

A similar example is:

9.c. {Orographic} or {relief rainfall}: [Moist air from the sea is forced to rise by the mountains. Cooling takes place, clouds form and rain falls on the windward slopes of the mountains. The lee slopes receive little rain because most of the water vapour originally in the air has fallen on the windward slopes].

Here the colon after the two technical terms can be replaced by 'is when':

Orographic or Relief Rainfall is when. . . .

Elaboration is a recursive system in the technicalizing process, so that one technical term can have two, three, or even more elaborations on a single term. It's quite common to find a sequence of elaborations, exploiting different grammatical structures, such as:

> 10. At [the lowest level], {tropic level 1} [(that is, where life forms are the simplest)], . . .

The means of elaboration of a technical term appear to vary according to the degree to which the phenomenon concerned is more or less accessible to the senses. Phenomena less accessible (e.g., transpiration, orographic rainfall) require more elaboration than tangible physical phenomena such as mesas or buttes.

The elaboration of technical terms does not only occur within textbooks. As the following examples show, teachers often provide oral elaborations of technical terms in class (recorded by P. Wignell):

> 11. They are {subsistence farmers}, which means [they only grow enough for themselves].
> 12. {Rainfall effectiveness}, which is [how effective the rainfall is], this depends on {seasonality}, which is [what season of the year the rain falls in], and {intensity}, which is [how hard the rain is], really hard rain in a short time is not as effective as the same amount of rain over a longer period because with really heavy rain most of it runs off and isn't absorbed.

Although elaboration is by far the most common means of defining technical terms, there is one type of definition which does not use the token/value structure. This is when a phenomenon is defined through a possessive attributive process: that is, a phenomenon is defined by an accumulation of attributes. For example, a desert can be identified as the sole possessor of all the following attributes:

> 13. (A desert has):
>   — lack of water,
>   — very low humidity,
>   — long drought followed by deluge,
>   — minimum cloud and maximum sunshine,
>   — very hot days and very cold nights,
>   — a great daily range of temperatures

This definition is descriptive and looks rather like the pre-Linnaean description of a plant, where something becomes identifiable by having a particular set of attributes. Each of the attributes is a necessary condition for a desert, but only the whole set is sufficient to define the term. This kind of definition tends to be used when the major classes in taxonomies are being established. Geographers look first for partial likenesses to establish principal groupings; subclassification then proceeds on the basis of partial difference. Text 3 is a similar example from science.

TEXT 3

Living things show — movement; responsiveness; assimilation; growth with development; reproduction. If you examine an object and it shows all of these properties, then the object is living. (Messel, Crocker, and Barker, 1964, p. 82)

Ordering the Experiential World: Technical Taxonomies

It was suggested earlier that setting up a technical vocabulary was geography's way of observing and describing the experiential world. It is only once phenomena have been named that the geographer's second task, grouping and classifying, can proceed. This involves relating the technical terms to each other in a systematic way through taxonomies, even though, as mentioned before, these taxonomies may be largely implicit. It is far more common for a science textbook to include a labelled diagram summarizing taxonomic relations than for a geography one.

Language embodies a number of lexical and grammatical resources for creating the taxonomic relationships of superordination (*a* is a kind of *b*) and meronymy (*a* is a part of *b*). In the geography texts the main grammatical resources used to realize these taxonomic relationships are relational processes and nominal groups (for analyses used here, see Halliday, 1985*a*).

Relational Processes

Intensive attributive relational processes (e.g., physical geography is a science; jellies, gelatin, and agar, are an unusual group of colloids which are solid or semisolid) are rare. This is because this clause structure classifies: It assigns a subclass to a class, but it is not semantically invertible, and so cannot be used to sub-classify (to break down a class into its sub-classes).

14.	Physical geography	is	a science.
	Carrier	Process:	Attribute
	(sub-class)	intensive	(class)

Possessive attributives, on the other hand, do move from wholes to parts, and so are naturally used to decompose. Somewhat metaphorically, this same structure is used to break classes into sub-classes. This means that geographers generally talk about relationships in their superordination taxonomies in terms of parts and wholes, with the meaning: This term is part of the taxonomy. For example,

15.	Desert landforms	consist mainly of	those due to erosion and those due to deposition
	Carrier	Process:	Attribute
	(class)	possessive	(sub-classes)

In Table 8.3 are some further examples of attributive relational processes. The fact that the same clause structure, the possessive attributive, is used both to

Table 8.3: *Examples of Attributive Relational Process*

(a) Relational processes of attribution which classify:
   a is a *b* (superordination):
   a. The rainforest is an ecosystem.
   b. Desert vegetation consists of four groups: . . .
   c. Desert landforms consist mainly of those due to erosion and those due to deposition.
   d. Precipitation includes rain, fog, sleet, hail and snow.

(b) Attributive relational processes of possession which have the function of decomposing:
   a has a *b*, *a* includes a *b* (composition).
   a. Every ecosystem has 2 parts: the physical environment and the biome.
   b. Each member of the ecosystem, no matter how small, has its own special place in nature's scheme . . .
   c. (the physical environment) includes all those things which provide energy, raw materials and life support for the community . . .
   d. (the biome) includes all the resident species of plants and animals
   e. Rainforests consist of many . . .

sub-classify and decompose brings out the essential continuity between superordination and composition taxonomies. However it is also a potential source of confusion for students who may not pick up on the different kinds of relationships being constructed.

Nominal Group Structures

Relevant nominal group structures indude:

   a. Pre-Classifier: realized as types of, kinds of, sorts of (realizing hyponomy), (e.g., types of environment, types of physical environment, types of rainfall).
   b. Pre-Deictic: realized as parts of, elements of, aspects of (realizing meronymy), (e.g., elements of the environment, parts of the ecosystem).
   c. Classifier^Thing: realizing superordination; once a technical term has been set up it is then open to sub-classification through the addition of a Classifier element; for example, once you have defined biome and environment you can have biotic environment which contrasts with physical environment, (e.g., orographic rain, frontal rain, convectional rain).
   d. Possessive-Deictic^Thing structures realizing meronymy (e.g., the rainforest's canopy).

These nominal group structures are often used in existential and material processes to sub-classify and decompose. For example:

16. *Existential process*
There are a number of other types of colloid. (Heffernan and Learmonth, 1981, p. 98).

17. *Material process*
Three distinct types of particle may be obtained from atoms: 1. The Electron . . . 2. The Proton . . . 3. The Neutron . . . (Heading, Provis, Scott, Smith and Smith, 1967, p. 5).

Geographers spend a great deal of time constructing taxonomies for textbooks, although this is never explicitly stated (nor backed up by some kind of two-dimensional diagram such as those presented below). For example, in Text 4, a variety of grammatical and lexical resources being used to present partial superordination and composition taxonomies of the ecosystem is displayed. Technical terms are introduced, elaborated, and then ordered taxonomically.

TEXT 4

What is an ecosystem?
The surface of our planet, where earth, air and water meet and where plants and animals (including man) live, is known as the biosphere. The biosphere is made up of many hundreds of different ecosystems. The word ecosystem comes from the Greek word '*oikos*' meaning 'home' and an ecosystem is that home or place in which a community or group of interacting plants and animals lives.
Every ecosystem has two parts: the physical environment and the biome. The physical environment provides the home for the living residents; it includes all those things which provide energy, raw materials and life support for the community such as the sun, climate, landforms and soil. The biome is the living part of the ecosystem and it includes all the resident species of plants and animals . . . There are two forms of life in the biome — plants and animals — and the basic difference between them is nutrition. Plants are able to make their own food using the sun's energy in a process called photosynthesis, while animals have to eat ready-made food.
Every living thing is dependent on both its physical environment and upon other members of its biotic community or biome. Also, in time, each living thing affects the environment for all other living things until eventually all the species of plants and animals in the ecosystem live in a balanced state — this means that they depend upon one another and live in harmony unless disturbed.
Each member of the ecosystem, no matter how small, has its own special place in nature's scheme . . . At the lowest level, trophic level 1 (that is, where life forms are the simplest), green plants convert the energy of the sun into food (carbohydrates and protein) which is stored in leaves. Thus green plants are called producers . . . At the next level (trophic level 2) animals eat these plants (herbivores) or eat other animals (carnivores); these animals are the consumers of the ecosystem. At the final level (trophic level 3) bacteria, fungi and certain protozoa (single-celled animals) break down dead animals and plants as well as the waste products of the consumers. These they convert into useful elements which are taken up once more by the roots of plants and so they are called converters. (Sale *et al.*, 1980, pp. 12–13)

Figure 8.7: *Composition Taxonomy of the Ecosystem*

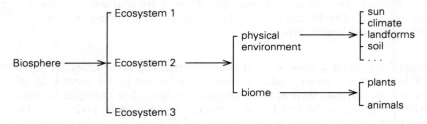

Figure 8.8: *Superordination Taxonomy of the Ecosystem*

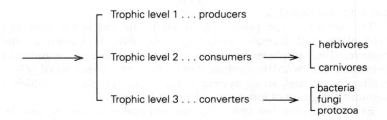

The composition taxonomy derived from this text is shown in Figure 8.7. A superordination taxonomy can also be derived from the same text (see Figure 8.8). Figures 8.7 and 8.8 demonstrate that taxonomizing in geography is implicit: The taxonomies are not presented formally to students — it is their task to retrieve them from the text.

Explaining the Experiential World: Implication Sequences

After grouping and classifying the experiential world, the third task of geography is to analyse and explain how phenomena come to be the way they are.

It has been shown that the most common way of introducing technical terms, prior to positioning them in taxonomies, is via an identifying relational clause, usually of the type 'An *x* (technical term) is a *y* (definition)'.

However, technical terms may also be defined by a relational identifying clause where the value is an anaphoric referent for a preceding descriptive sequence of text, which may be a clause or a series of clause complexes (for example, see Text 5).

Text 5

17. After flash floods, desert streams flowing from upland areas carry heavy loads of silt, sand and rock fragments. As they reach the flatter area of desert basins, they lose speed and their waters may also soak quickly into the basin floor. The streams then drop their loads, the heaviest materials first — the stones — then the sand and finally the silt.

Choked by their own deposits, these short lived streams frequently divide into a maze of channels spreading their load in all directions. In time, fan or cone shaped deposits of gravel, sand, silt and clay are formed around each valley or canyon outlet. These are called alluvial fans. (Sale *et al.*, 1980, p. 54)

The kind of definition in Text 5 introduces the notion of causal relations among phenomena. In elaborating the technical term one not only says what a thing is, but how it got to be the way it is: the identification of a phenomenon preceded by an explanation.

The kind of definition used for alluvial fans still involves a form of elaboration. However, in the clause complexes which describe the phenomenon the logico-semantic relation switches from elaboration to enhancement and, to a lesser extent, extension because the text records the temporal sequence of events by which the alluvial fan is created. (For discussion of logico-semantic relations, see Chapter 7 in Halliday, 1985a). However, the actual term, alluvial fan, is still defined by a relational clause with the technical term as the token. The referent 'these' summarizes the activity sequences and in its clause switches the relationship back to one of elaboration.

Where taxonomizing tends to focus on things (i.e., the naming and ordering of things), explaining tends to focus on processes. The emphasis shifts from things in place to things in action. To explain how things are, or came to be the way they are, it is necessary to use processes, participants, and circumstances. These tend to be arranged in clause complexes which will be called implication sequences here. This term will be preferred to Martin's (1984a, p. 5) term 'activity sequences', because it captures the fact that each step through the sequence implies what has gone before.

For example, if you take rain (i.e., precipitation), the fact that it is raining implies that there has been evaporation, uplift, adiabatic cooling, and condensation. Or, to go to another field altogether, dog showing, the fact that you are showing a dog implies that you have bought it (or bred it), fed it, groomed it, and trained it.

The term implication sequence thus indicates a more ordered connection among sequences than the term activity sequences. For example, take the short activity sequence: We went to the zoo and saw the tigers, then we saw the chimps, then we had lunch, then we saw some more animals and went home. The fact that we saw the chimps in no way implies that we saw the tigers. Here is an example of an implication sequence taken from seventh grade (12–13-year-old) class notes.

18. Frontal uplift occurs when cold air meets warm air forcing the warm air to rise.

In this example the technical term 'frontal uplift' is defined by enhancement: that is, it is being defined by its cause. It is printed in the notebook with the technical term in capitals and underlined, followed by its implication sequence. The technical term is thus presented as the name of the implication sequence.

The implication sequence for convective uplift is as follows:

19. CONVECTIVE UPLIFT
Air in contact with a warm surface will become heated and expand causing it to rise. Dew point will be reached, condensation will take place,
and convectional clouds will form.

In both these examples (frontal uplift and convective uplift), the technical term
has a Classifier^Thing structure. The general term 'uplift' is a nominalization of a
process and the classifier is what would be the agent in a clause, that is the thing
that is forcing the air to rise. Thus, when the process and medium are conflated
into the thing in a nominalization, agency comes out in the classifier.

The last three clauses in the last example do not define. Rather, they take
technical terms which have already been defined and arrange them in a sequence
of events in which each event implies the one or ones before it; that is, clouds do
not form until condensation has taken place, condensation does not take place
until dew point has been reached, and dew point is not reached until air containing water vapour has risen and cooled. It is through these implication sequences
and combinations of them that 'explaining' is achieved in geography.

Modelling Implication Sequences

Although it is quite easy to model taxonomies because there are conventional
models available courtesy of the natural sciences, the modelling of implication
sequences is more problematic. They represent a more dynamic ordering of the
world than taxonomies. Things happen to other things in particular environments
over time which then cause other things to be or to happen. A way to capture this
is through the use of transition networks (for a detailed discussion of transition
networks, see Winograd, 1983).

In short, the basic components of transition networks are states (represented
by circles) and arcs (represented by arcs with arrows representing the direction of
change). An arc represents the movement from one state to another. A state
represented by concentric circles is a terminal state. This is a place where the
sequence either does, or could, stop. As well as being a potential terminating
point for that sequence, it is also a potential starting state for another sequence.

By isolating the basic linguistic categories found in implication sequences
and superimposing them on transition networks, they can be modelled quite
efficiently. The basic linguistic categories involved are processes, mediums,
agents, and circumstances (see Halliday, 1985a) plus, what might be termed
results. The first four of these are superimposed on the arcs, and results are placed
in states.

A generalized model is shown in Figure 8.9. In Figure 8.9, State B, represented by concentric circles, would by the terminal state for this implication
sequence and a potential starting state for another sequence. The elements in
parenthesis are optional. The only obligatory elements on an arc are a Process +
Medium.

Arcs could also be labelled with nominalized processes, which would, in effect, conflate the roles of process and medium. For example, a non-nominalized
way of expressing condensation is: water vapour condenses. In this nuclear structure, Medium + Process, condensation is represented as a process (there is

Figure 8.9: A Generalized Model

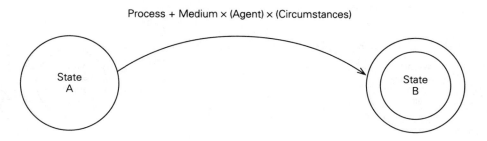

Process + Medium × (Agent) × (Circumstances)

some condensing going on), and water vapour as the medium through which it is going on. In the textbooks this is usually realized as 'condensation occurs', which is again a Medium + Process structure. The original medium, water vapour, rather than being omitted, has been absorbed into the technical nominalization, and the dummy process 'occurs' replaces the original process 'condenses'.

It is interesting to note that although terms such as evaporation and condensation are nominalizations of processes, in geography books they are still regarded as being processes; for example, 'condensation is a continuous process'. The role of the dummy process here appears to be that once a process has been made into a thing, it loses its capacity for action. So, to be put back into action, a dummy process must accompany it.

The reason for this curious cycle of turning processes into things, and then finding a way of turning them back into verbs again, is that nominal group resources in English allow for the possibility of classification, qualification, and description, whereas the verbal group resources do not. In order to be classified and described, processes must be made into a thing, even though, for all intents and purposes, scientists still conceive of them as processes, and commonly refer to them as happening, occurring, taking place, and so on.

These process nominalizations deriving from implication sequences are difficult to interpret in taxonomic terms. They are however technical terms. What appears to be happening is that rather than being used to taxonomize, they are being used to technicalize implication sequences or parts of them.

Figures 8.10–8.15, show an example of modelling implication sequences. Weather is the topic used, as taught to seventh-graders, with some reference to an undergraduate textbook to fill some gaps.

Jargon or Shorthand: The Functions of Technicality

Sketched out above is what could be called the technology of geography; that is, how geography takes everyday things, renames (and by doing so reorders), and explains them. This could be described as a technology of technicality.

The linguistic realizations of technicality are related to the register variable of field. Field refers to how the experiential world is divided up into institutional areas of activity, such as dog showing, sailing, linguistics, botany, and so forth.

Figure 8.10: *How Water Gets into the Air*

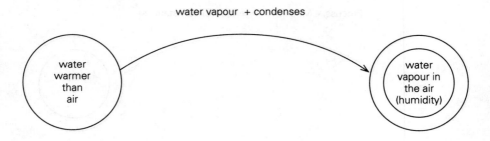

water vapour + condenses

Figure 8.11: *How Clouds Form*

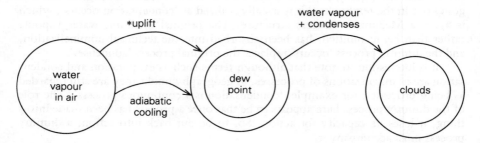

*uplift = (air + rises × Agent)
Agent = mountains, front or convection
[two arcs = simultaneous processes]

Figure 8.12: *How Rain Falls*

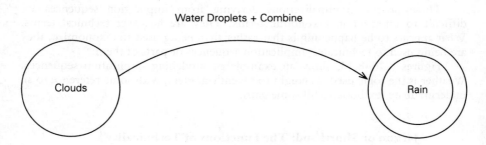

Water Droplets + Combine

Field is closely linked to experiential meaning in the grammar, and is realized through patterns of transitivity and lexis (for further discussion, see Halliday and Hasan, 1985).

It is a characteristic of all fields that they name the things concerning them. Therefore each field develops its own vocabulary, and from looking at the lexis

Figure 8.13: How Cloud Droplets Combine—A

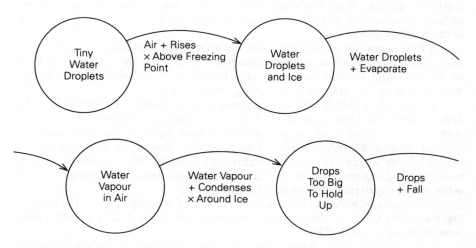

Figure 8.14: How Cloud Droplets Combine—B

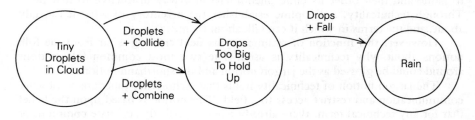

Figure 8.15: Adiabatic Cooling

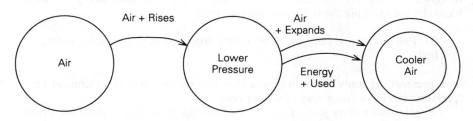

used in a text, its field can usually be identified. For example, a text containing the words ring, judge, terriers, handlers, dog bitch, and grooming is likely to be from the field of dog showing.

However, fields not only name the things that interest them. They also order those things taxonomically. It has already been demonstrated how the grammar has resources for creating taxonomic relationships among things. For example, a text containing the words terriers, Staffordshire bull terriers, fox terriers, Yorkshire terrier, West Highland White, and so on, is drawing on a super-ordination taxonomy of terriers realized through the Classifier^Thing structure of the nominal group.

A field is not just a collection of things related taxonomically. It is also a set of related activities: that is, what the things in the field do. For example, the field of dog showing implies sequences of activities such as breeding, grooming, nurturing, showing, judging, prize giving, and the like.

The extent to which one can be considered an insider of a particular field depends upon the knowledge of the lexis, taxonomies, and activity sequences it contains. For example, given the terms backwash squeeze, end play, dummy reversal, double dummy, duck, and turkey, one's control of this field can be judged by one's ability to use these terms appropriately. To be an insider means understanding the meaning of the terms, their taxonomic relationships to each other, and the activities that the field involves.

The degree of technical lexis in the discourse of geography raises the following questions: What are the functions of technicality in text? Why would a field want to set up a technical lexis?

It has already been shown that the major function of technicalizing in geography is to enable the setting up of field-specific taxonomies. In this respect, then, technicality has a field-creating function: It is the resource a discipline uses to name and then order its emic phenomena in a way distinctive to that field. Through technicality, a discipline establishes the inventory of what it can talk about, and the terms in which it can talk about them.

However, this function of technicality as field creating is far from the folk notions about why technicality is used. The two most common folk notions around could be glossed as the jargon notion and the shorthand notion.

The jargon notion of technicality holds that technicality functions to obscure communication and restrict access to a field. This notion is based upon the belief that for any technical term, there already exists a perfectly adequate common, or garden variety word, which could be used in its place, so that technicality is just a process of substituting names for things, and that the use of the jargon word is designed to impress listeners or to create feelings of inadequacy in them. Consider for example the following student text:

20. Precipitation is only moisture falling from the sky, e.g., rain, snow, hail.

The presence of 'only' here indicates that the student sees the technical term, 'precipitation', as a fancy way of saying rain.

If technicality were just a fancy way of talking about things there are already names for, it would amount to nothing more than jargon. However, two pieces of evidence form the grammar conflict with the jargon notion of technicality and provide support for claiming that technicality is not just a translation of everyday phenomena into geography-speak.

The first is the token/value structure, and the second is the use of the taxonomizing resources of the grammar. It was noted earlier that technical terms realize token, and definitions realize value. This means that a model for technicality is closest to one of translation. Saying that, in French, *chien* means dog, is the same as saying in geography, ecosystem means the place where . . . And, just as in interlanguage translation there is no simple substitution of terms, so in geography one is not just giving a fancy name to a phenomena already known as something else. Consider the famous example of *savoir* versus *connaître* in French, which both translate into English as 'know', but, which enter into a semantic opposition in French: *savoir* means 'to know (be aware of) something', connaître means 'to know (be acquainted with) someone'.

There is no difference between this and the contrast between a folk taxonomy of climate (humid, tropical, cold, dry) and a geographer's taxonomy of climate (tropical humid, subtropical, temperate, polar). In the folk taxonomy Brisbane's climate would be described as humid, whereas to a geographer it would be classified as subtropical. The folk taxonomy cannot be simply translated into geography, since geography sets up different oppositions, and to express these oppositions it needs to introduce and define its own terminology.

The token/value structures of the grammar provide the means for setting up the technical lexis, and the taxonomizing resources of the lexico-grammar provide the means of setting up the new oppositions among terms. Once the word 'climate' has been established, it becomes possible to base relationships of superordination and meronymy between 'climate' and other terms in the taxonomy.

Attaching a token to a value is giving the term a valeur in the system within that field. The discourse of geography consists of a set of technical terms which relate in certain taxonomic ways. The oppositions between the terms are specific to the field of geography. It is not that in geography one is necessarily creating new things, but that one is setting up field-specific oppositions between things.

The second less pejorative notion commonly held about the function of technicality is that it acts as a kind of shorthand: It enables a more concise communication among specialists.

The analysis of technical texts suggests that there is a good deal of truth in this notion, and that technicality does function to distill, to develop a shorthand way of talking about the world from a geographer's point of view.

This function of technicality to distill or condense is seen when we try to unpack technical texts; that is, try to rewrite them without using any technical terms. This procedure reveals that there are degrees of distillation.

1   If we just try to unpack the technical terms that are names for things, it's noticeable that what one does with single nominals (e.g., names of animals, landforms, etc.) is simply to substitute a folk name for the technical name. For example, mesa becomes hill. In these cases it is found that the technical term has not distilled or condensed information: we are substituting one word for another word. However, the technical term did provide a more delicate taxonomic classification which is lost by performing folk substitution. For example, by calling a mesa a hill, what differentiates a mesa from a butte is lost (a butte would also be called a hill in folk terms). In order to describe the difference between a mesa and a butte we would need to add descriptive epithets and qualifiers to hill: for example, a mesa is a fairly large hill with steep sides and a flat top, and a butte is a smaller hill with steep sides and a small flat top. In adding the epithets and qualifiers more information is gained about the terms,

but their classification in a taxonomy is lost. We are describing, not classifying. Distilling thus involves shedding the epithets and qualifiers and concentrating on classifiers and things.

2   If we unpack technical terms for things which are complex phenomena, where there is no one-to-one relationship between name and thing, such as ecosystem, the text becomes long and unwieldy as we substitute an often lengthy value for the token. In these cases, technicality is already showing dividends in terms of distilling content.

3   But it is when we try to unpack technical terms generated by implication sequences, that we realize how far technicality has distilled or condensed meanings. For example, the term 'adiabatic cooling' provides a shorthand for all of 21 (below).

21. As air is moved upward away from the land-water surface or downward toward it, very important changes occur in the air temperature. Air moving upward away from the surface comes under lower pressures because there is less weight of atmosphere upon it, so it stretches or expands. Air moving downward toward the surface from higher elevations encounters higher pressures and shrinks in volume. Even when there is no addition or withdrawal of heat from surrounding sources, the temperature of the upward or downward-moving air changes because of its expansion or contraction. This type of temperature change which results from internal processes alone is called adiabatic change. (Trewartha, 1968, p. 136)

If technicality consisted simply of reorganizing existing folk taxonomies, it could not be justified as a shorthand. But technicalizing processes does save space. The implication sequences discussed earlier generate technical terms, either through the nominalization of the process or through the naming of final states, and so on. These do not become taxonomized in the text, but technicalizing the processes represents a significant gain in compactness. And the processes, once nominalized, could be taxonomized.

Two final observations can be made from attempts to unpack technical texts. The first is that unpacking presupposes knowledge of the field. That is, many of the terms cannot be unpacked without prior knowledge of how things are taxonomist in that field. This in turn presupposes that the reader has at some point learned the vernacular elaborations of such terms. For example, it is not possible to unpack the term biome without some knowledge of what an ecosystem is and where the biome fits into it.

The second observation to be made is that unpacking a technical text does not appear to make a lot of sense. It vastly increases text length and makes the amount of information to be processed relatively unmanageable.

It is, however, important to unpack a term the first time it appears. This is what gives the definition and allows the term to be used throughout the rest of the text. The evidence thus suggests that technicality is not meaningless jargon. Technicality functions as a field-creating process, allowing the setting up and taxonomizing of areas of human interest. The use of a technical lexis makes it possible to distill or compress meanings. One result of distillation is that those who share a particular field are saved the time-consuming process of continual elaboration and can get on with their primary concern: the observing, ordering, and explaining of new phenomena. Moreover, it seems that the more a field is concerned with explaining phenomena — rather than just ordering them — the greater the distillation offered by technicality.

Conclusion

The discipline of geography is concerned with making order and meaning of the experiential world, through observing, classifying, and explaining phenomena. It has been shown here that the discourse of geography observes the world by setting up a technical lexis; that it orders the world by arranging these terms into taxonomies; and that it explains the world through implication sequences of cause and effect. It has also been suggested here that in the process of technicalizing, a field is created: The set(s) of meaningful oppositions, which define a particular social institution, such as geography, are established.

The field of geography is thus made up of a number of interrelated taxonomies and sets of implication sequences, realized by technical terms. The major task of a geography textbook is to elaborate the technical taxonomy and generate terms for how things come about.

Learning the discourse of geography, learning to be a geographer, entails learning the technical terms and their valeur within the system of the field.

Geography teachers and textbooks are fond of emphasizing that geography is all about interrelationships. The linguistic evidence adds substance to this claim, for indeed much of geography is about the interrelationships between terms in taxonomies. However, while the natural sciences sometimes make their taxonomies explicit (see Shea, 1988), geography almost never takes this step. The taxonomies are there and are built up through the lexico-grammar but are not explicit, in that they are not displayed (in network diagrams such as those used to represent taxonomic relationships throughout this article). The relationship between terms has to be extracted from the text. Thus, the student has not only to find order and meaning in the experiential world, but also to uncover the order and meaning latent in the discourse of geography.

Chapter 9

Literacy in Science: Learning to Handle Text as Technology*

J.R. Martin

Evaluation

Early in 1988 Ben, age 8, wrote the following explanation of the origin of our planet:

Our Planet

Earth's core is as hotasthe (x hot as the) furthest outer layer of the sun. They are both 6000c°. Earth started as a ball of fire. Slowly it cooled. But it was still too hot for Life. Slowly water formed and then the (firstsx) first sign of life, microscopic cells. Then came trees. About seven thousand million years later came the first man.

[accompanied by a picture of the earth, with certain of what appear to be continents marked]

His teacher commented on his text as follows:

'Where is your margin? This is not a story'.

And on his illustration:

'Finish please.'

Ben's parents were naturally rather upset. Small wonder. Ben has a keen interest in science and enjoys reading in this area. His text is clearly influenced by his reading, but not copied word for word. And it is a perfectly readable account of the history of our planet. As a piece of scientific writing, taking the child's age into account, it is quite appropriate. Yet like many other young writers interested in the nature of things, Ben is being given a clear message: 'Literacy in science is not the business of primary school; here we write narratives. Drawing is important; spend more time on your pictures'. Or to put this in more blatantly sexist

* This chapter is taken from *Literacy for a Changing World*, 1990, Hawthorn, Victoria, Australia, ACER, chapter 4.

terms: 'Why don't you write like a girl?' (see White, 1986 for a discussion of science, gender and writing.)

It is obvious what has gone wrong: the teacher's expectations have not been met. More seriously, these expectations have not been made explicit to Ben. Instead of being evaluated on its own terms, as a perfectly valid piece of writing in science, it is rejected outright. Things could have been worse. The teacher might simply have given it a lower mark or more negative comment than a narrative, obscuring further the criteria on which the evaluation is based. As things stand, the message is clear, misguided though many of us might feel it to be.

The point of course is that writing in science is different from writing in other parts of the curriculum and that scientifically oriented criteria have to be used to evaluate it. Developing subject-specific criteria for evaluating writing is an urgent need in Australian schools. And this implies a much greater consciousness of the special nature of writing in different areas of the curriculum among teachers than is currently made available through pre-service training, syllabus documents and what little in-service training remains.

In order to develop scientifically oriented evaluation criteria we have to be very clear about the kind of knowledge science is trying to construct and also about the ways in which scientists package this knowledge into text. Neither the knowledge nor the packaging can be understood without looking very closely at the language scientists work with when they talk and write. The language science uses to construct scientific understandings will be explored; the written genres used to document and explore this knowledge will be reviewed. Lemke (1990a) provides a useful introduction to the social semiotic perspective on the role of language in science adopted in this chapter.

Adopting such a perspective is critical at this time because of the steady emergence in Australia of documents stressing the importance of using non-scientific language to explore science in primary and junior-secondary schools. The recent NSW *Writing K-12* document for example heads its list of recommended forms of writing in science with stories, especially science fiction, personal reports and descriptions of observations, plays, poems and cartoons (1987, p. 125). The same department thrusts students into very demanding forms of scientific writing in years 11 and 12 and uses writing of this kind as the basis for evaluation in the state-wide High School Certificate examination. Christie (1989, section 1) includes a valuable case study of one student's experiences coping with the literacy demands of senior-secondary physics and environmental science in Victoria, for which he had received little preparation in the previous ten years of schooling. As more and more emphasis is placed on doing science 'in your own words' (Wignell, 1987) the situation is deteriorating rapidly, to the point where young science writers such as Ben are not only not encouraged but are actually being nipped in the bud. Without a clear understanding of the fundamental role of scientific language in doing science, this problem cannot be properly redressed.

I will make use throughout of quotations from a variety of science textbooks along with examples of student writing. The textbooks are important because they are the main source of models of written scientific language for most students; they are also focal because most extended writing in science is in fact copied more or less directly from such books (see Wignell, 1987). The reason for this is that writing in science is not taught, and students have no better way to learn. Whatever their demerits, the removal of traditional textbooks (e.g., Messel *et al.*, 1964) from science classrooms over the past twenty years has meant that increasing numbers of students are exposed to fewer and fewer models of scientific discourse. This is compounded by the fact that recent textbooks (e.g., Cull and Comino's, 1987) provide very fragmented models of scientific text. Both developments are fundamental obstacles to learning the language of science in Australian schools.

Literacy in Science

One of the first images that comes to mind when thinking about science is the experimental laboratory. People in white coats are busy with various sorts of technological apparatus — scientists with the tools of their trade. Most of us are also aware that scientists can be hard to understand. It is not only their tools which are technical; their words are technical too. There are not many people in our culture that can make sense of the following text:

Coniopteris fruitformis n. sp.

DIAGNOSIS

Leaf, bi- or tri-pinnate, length unknown, portions preserved maximum length 15 cm. Pinnae, length 1–10 cm., attachment alternate about 1 cm apart at angle of about 60 on rachis. Pinna shape tapering arrow tailed, apex pointed. (Heffernan and Learmonth, 1981, p. 221)

And it is not just the words. The grammar is special too. The text is not written in sentences, but in long nominal groups (or naming groups, such as **the text**, **sentences** and **long nominal groups** in this sentence). The point of both the technical terms and the grammar is to compress as much information as possible into a short space. The palaeontologist at work wants to document the fossil as economically as possible — there are many thousands of other fossils to record.

We can see from texts such as this that one of the most important pieces of technology used by scientists, and one that is often overlooked, is language. To be literate in science means to be able to understand the technical language that is used. To understand this we have to look more closely at what scientists are trying to do.

Science and Common Sense

Scientists think about the world differently from other people. At times they are very critical of common sense understandings:

Meteors and meteorites

Meteors, often wrongly called 'shooting stars' or 'falling stars', look spectacular as they flash briefly across the night sky (Figure 2.42). They, however, have nothing to do with stars. They are simply tiny pieces of rock rushing through space and being burned up by friction as they hurtle into Earth's atmosphere at speeds of above thirty kilometers per second. (Heffernan and Learmonth, 1981, p. 34)

Here the astronomer is criticizing common sense for classifying meteors as a kind of star. She sees the world differently, defining meteors as **tiny pieces of rock rushing through space and being burned up by friction as they hurtle into Earth's atmosphere**. Underlying this is a picture of life, the universe and everything in which the universe is composed of billions of galaxies, and galaxies are composed of millions of solar systems. Solar systems are organized into parts and wholes along the following lines:

Figure 9.1:   Organization of Solar Systems: A Scientist's View

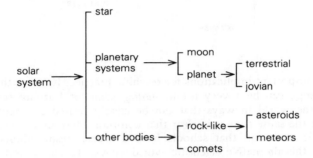

In this world view meteors are only distantly related to stars. But we should not be too quick to dismiss common-sense understandings. To the naked eye, stars, meteors and planets look very much alike; and a meteor is indeed a star that falls from the sky. The common sense view of heavenly bodies can in fact be mapped out in a similar way to the scientist's, though the relationships are very different. In particular one's point of departure is the planet Earth, rather than the universe. Looking up, we see what Figure 9.2 represents. When we compare the common sense and scientific maps of the universe we can see that both organize reality. This they have in common. What differs is the criteria used as the basis of this organization. To the naked eye the morning and evening stars are different objects and quite justifiably stars: small shining objects that appear in the sky at dawn and dusk. For science, they are the same planet — Venus: looked at through a telescope in an observatory it is clear why a scientist adopts this point of view.

The common sense view depends on careful observation with the naked eye. Science augments this with observatories, radio telescopes, space ships, studies of meteorites (meteors which make it to the ground) and other types of information gathered in various ways and accordingly produces a different picture. Rather than saying that science is right and common sense wrong (or vice versa when religious beliefs are introduced as evidence), it is more important to understand common sense and science as different pictures of reality, based on different organizing criteria. The function of science then is to construct an alternative interpretation of our world to that provided by common sense. In our culture, this is its job.

Figure 9.2: Organization of the Universe: A Common-Sense View

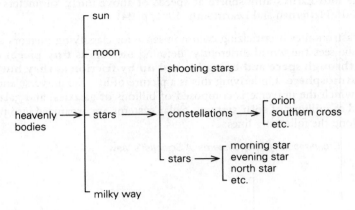

This has important implications for teaching practice. It means that common sense knowledge can be a very useful *starting point* for learning science, since it organizes the world in ways that can be clearly related to scientific understandings. At the same time it is clear that common-sense understandings differ from scientific ones and that schools have a crucial responsibility to induct students into the alternative scientific world views. Teachers need to be constantly aware of the dangers of stranding students in their own words. This guiding role, bridging across common sense and science, is put very clearly by Britton.

> Surely it is the links between 'commonsense' and 'theoretical' concepts, the links between 'ordinary language' and 'theoretical language' that make learning possible — whether in school or out — and it is the ability to move back and forth across that continuum that characterizes thinking at any mature stage. (Britton, 1979)

Organizing the World: Things

Classification

Classification is a fundamental part of every science. One important function of science textbooks is to introduce novices to this form of organization as it has been projected onto reality as a result of centuries of research. The following text is designed to socialize high-school students into one aspect of chemistry — mixtures:

> **Mixtures** are substances that can be easily separated without making any new chemicals. We have already seen that most of the materials around us are mixtures (Chapter 3). The air, water, soil and rocks around us are all made of chemicals mixed together. Mixtures can be divided into a number of groups (Figure 5.1).

Solutions are mixtures that have the same properties throughout. When sugar is mixed with water the solid crystals disappear. The solution formed can be seen through and looks the same throughout. If we mix sand with water, we see that the *large* grains soon settle to the bottom. Here it is easy to see solid and liquid mixed together. If soil is mixed with water, some of the fine grains, which we can just see, take a long while to settle to the bottom. A **suspension** is said to have been formed if these fine grains can also be filtered out. When some clay is mixed with water containing a little sodium hydroxide, a mixture with properties in between solutions and suspensions is formed. The very tiny grains do not settle out but do pass through filter paper. Such a mixture is called a **colloid**. (Heffernan and Learmonth, 1981, p. 90)

This passage is organizing mixtures into three sub-types. The relationships can be represented in a diagram like those used above, where terms to the right sub-classify those immediately to their left:

$$\text{mixture} \rightarrow \begin{cases} \text{solution} \\ \text{suspension} \\ \text{colloid} \end{cases}$$

Each term in the diagram is a technical term. And in the textbook, these terms were specially highlighted in bold face. At a first glance, three of the terms are familiar every day words: **mixture, solution** and **suspension**. One is unfamiliar: **colloid**. This term is used mainly in chemistry, and can thus be described as indexical of the field (i.e., once we hear the term, we know what field we're in). It is this kind of term that is commonly referred to as *jargon*.

We must however be cautious here, because the familiarity of **mixture, suspension** and **solution** is deceiving. None of these terms mean the same thing in chemistry as they do in common sense. They as just as much jargon as **colloid**. The text makes this very clear by defining each term, giving it its special meaning in chemistry:

Term	Definition
mixture	substance that can be easily separated without making any new chemicals
solution	mixture that has the same properties throughout
suspension	mixture in which fine grains can be filtered out
colloid	mixture in which tiny grains do not settle but do pass through filter paper

Science typically uses a special kind of clause to define technical terms — the identifying relational clause (clauses that explain who's who and what's what). These clauses are always reversible; and the technical term is Subject (coming first in the clause) when the voice is active (for a discussion of relational clauses see Halliday 1985a; Chapter 8):

active
Solutions are mixtures that have the same properties throughout.

passive
Mixtures that have the same properties throughout are (called) **solutions**.

The function of definitions is to translate common sense understandings into scientific ones. This always involves a certain amount of condensation. It is easier to refer to **mixtures** than to **substances that can be easily separated without making any new chemicals**. Without this condensation scientific texts would become very long, and probably unreadable, even for professionals.

The process of translating common sense into science is a long one. Most definitions use familiar technical terms to introduce new ones (e.g., **substances** and **chemicals** in the definition of **mixtures** above). So only a small minority of scientific terms are ever completely defined in every day words. Alongside defining technical terms, the passage also organizes these terms with respect to each other. The critical sentence is **Mixtures can be divided into a number of groups**. This tells us that solutions, suspensions and colloids are kinds of mixtures, enabling us to build up the diagram above. So knowing the definitions of the terms is not enough to understand the text; it is also crucial to determine precisely how the terms are related to each other.

Perhaps the best metaphor for technical language is that of **distillation**. Technical language both *compacts* and *changes the nature* of every day words — just as a vat of whisky is both less voluminous and different in kind from the ingredients that went to make it up. Marsupials for example are not simply Australian animals — wombats, possums and kangaroos. For the biologist they are warm-blooded mammals that give birth to live young with no placental attachment and carry the young in a pouch until they are weaned; and they contrast with the two other groups of mammals, monotremes (egg-laying) and placentals.

People sometimes complain that science uses too much technical language, which they refer to pejoratively as 'jargon'. They complain because the jargon excludes: it makes science hard to understand. This is a problem. Jargon is often used where it is not needed. And translating jargon into common sense is an important social responsibility of all scientists, which needs to be more seriously addressed. However, the simple fact is that no scientist could do his or her job without technical discourse. Not only is it compact and therefore efficient, but most importantly it codes an alternative perspective on reality to common sense, a perspective accumulated over centuries of scientific inquiry. It constructs the world in a different way. Science could not be science without deploying technical discourse as a fundamental tool. It is thus very worrying when syllabus documents discourage teachers from using technical language with students, especially in the early years (emphasis added):

Developing Science Vocabulary

Appropriate science vocabulary will become part of the child's spoken language. This vocabulary will be introduced *only when it is needed*, or when it is desirable to communicate more effectively about something that has *already been thoroughly explored* by the child. (Victorian Primary School Science Syllabus, 1981, Section 3.41)

The idea that personal writing can be used to do science is equally troubling. Given appropriate contexts for writing, children take up scientific language without difficulty at an early age (see Christie, 1986). Note how patronizing the following advice would be if given to a child like Ben, our young writer in section 1 and how misguided the unsubstantiated comment on using personal writing for abstract thinking:

Narrative and imaginative writing

9.2 Personal writing can be especially appropriate for the earlier years and the more concrete thinking levels. However, in certain contexts, it can be *appropriate for abstract thinking* as well. (NSW, *Writing K–12* Syllabus, 1987, p. 127)

Composition

Alongside classifying the world, science also reorganizes the world in terms of composition — the ways in which parts are related to wholes. One way in which a biologist does this is by using microscopes. This technology, combined with the use of various stains, has enabled accounts of cell structure like the following (note how the technology augments our senses in such a way that the scientist makes observations about phenomena for which there is no common sense classification):

All animal cells have a number of parts in common (Figure 9.9). They all have a **cell membrane**. This is a thin 'sack' that controls the chemicals that can enter and leave a cell. The liquid contents of a cell are called the **protoplasm**. These liquid contents are divided into the nucleus and cytoplasm.

The **cytoplasm** looks like a liquid that has very few other things in it. But it is really very complex. The cytoplasm is mostly water with many chemicals present. Most of the cell's chemical reactions take place in the cytoplasm . . . When a cell is stained many more things can be seen in the cytoplasm. A number of small 'sacks' called **vacuoles** are found. These store food materials and many other chemicals. Small dark rod-shaped objects called **mitochondria** can also just be seen. These move around the cell to provide energy where it is needed. (Heffernan and Learmonth, 1981, p. 152)

Here we find the same pattern of highlighted technical terms and definitions as with classification. What is different is the relationships among the terms. These are now of part-to-whole rather than class-to-sub-class. Sentences like the following control the composition:

Figure 9.3:   Composition Diagram

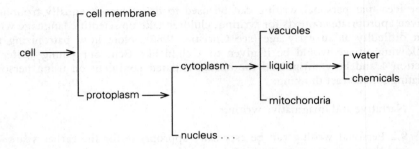

Figure 9.4:   Classification Diagram

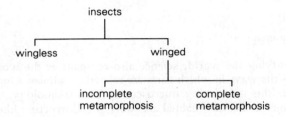

Source: Messel et al. (1964) Science for High School Students

All animal cells **have a number of parts** in common.
They all **have** a cell membrane.
The liquid **contents of** a cell are called the protoplasm.

The typical clause type used to establish the part–whole relations is possessive, with the verb **have**; these clauses are not reversible:

They all have a membrane.

But not:

*A membrane is had by them all.

The relationships built up can be diagrammed. See Figure 9.3.

Diagramming

In general science uses different types of diagram for classification and composition. The diagrams we have used so far are modelled on those used by systemic linguists. When classifying, a scientist would normally turn these diagrams around 90°, with the largest class considered on the top (Figure 9.4) (Messel *et al.*,

Figure 9.5: The Structure of an Insect. The wings on the left side have been removed.

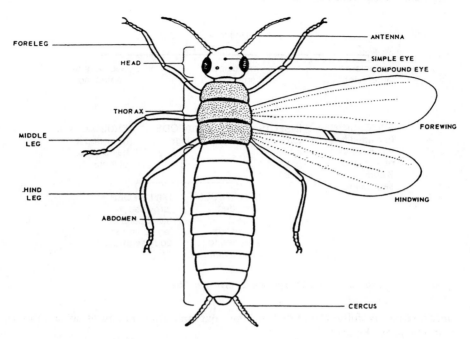

Source: Messel *et al.* (1964) *Science for High-School Students*.

1964). Diagrams of this kind model what are known as taxonomic relations and are referred to as *taxonomies*. Composition is handled through labelled pictures or diagrams. Messel *et al.*'s labelled diagram for the structure of an insect is as in Figure 9.5. When compositional relationships become complex, with parts, sub-parts, sub-sub-parts and so on, the kind of diagram used by linguists can be used to represent the relationships involved more clearly than a labelled picture.

One of the problems with all the diagrams used so far is that although they clearly display the relationships among the terms, they do not include an account of the criteria forming the basis of the organization. These have to be recovered from passages of text. This can be overcome to a limited extent by adding extra labelling to the diagrams. Cull and Comino's (1987) classification of arthropods is a diagram of the kind shown in Figure 9.6. Diagrams of this kind quickly become overloaded with information; and the explanations given are rather cryptic. But as a summary device they are no doubt very useful. Unfortunately, Cull and Comino use diagrams of this kind *in place of* text defining terms and explaining relationships among them. Textbooks like this are really little more than supplements to teacher explanations and xeroxed notes; they cannot function on their own as resources of science information.

Messel *et al.* (1964) amplify their diagram with examples of the insects in each class and pictorial representations of the criteria used. In their case this

Figure 9.6:   *Classification of Arthropods*

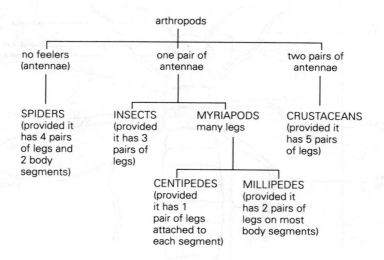

Source: Cull and Comino (1987) *Science for Living*, p. 162.

information is fully elaborated in the text and thus functions as a kind of summary. See Figure 9.7.

Morris and Stewart-Dore 1984, pp. 48–56 use the term 'structured over-view' to refer to diagrams displaying relationships of classification and composition (among others) in text. They recommend that these be used as an 'advance organiser to introduce readers to a topic' (1984, p. 53). Their appendix provides sample lessons in science, describing step by step how teachers might construct these with students (1984, pp. 201–202). Being able to shunt in this way between text and diagrams is certainly a fundamental aspect of science literacy. Morris and Stewart-Dore introduce the technique with a view to effective reading in the content areas (ERICA); from the perspective of writing, translating structured overviews into text would be equally useful.

For example, in preparing students to write science, teachers can work with them to build taxonomies of relevant scientific information on the board, discussing with them the kinds of relationships between the phenomena being considered. Here, a great deal of teacher-guided talk, in which the students re-hearse and clarify their understanding of such relationships will be an important part of preparing for writing. A subsequent step will involve beginning to plot the overall pattern of the scientific genre to be written. At this point teachers will need to prepare students for some of the linguistic features of the written genre, and, depending on their previous experience, considerable care will need to go into the examination of the genre to be produced. This is critical because of the differences between talking about and writing science (for differences between speech and writing see Hammond, this volume).

Teachers may in addition provide models of such genres on charts or in textbooks (McNamara, 1989, discusses the use of models by a secondary science

Figure 9.7:   Types of Insects

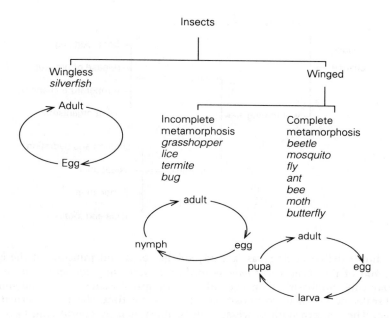

Source: Messel *et al.* (1964) *Science for High School Students.*

teacher), or, alternatively, they may write sample models on the board, involving students in helping with the writing. In either case, teachers and students should discuss the sequence of steps to follow, and some of the principal linguistic features which will mark the opening and the development of each phase in the text. Some possible ways to approach the teaching of writing along these lines are provided in Christie and Rothery (1989).

Once a repertoire of different scientific texts has been built up over several months of work, the nature of the discussion of the written genre for any particular science lesson will of course be considerably enriched, because students can then take all their accumulated experience of earlier writing, and bring it to bear on the new task. As they develop confidence in writing science, they will be able to adapt the writing to their own purposes, even to the point of spoofing it as current writing syllabuses are so eager to recommend. The point is however that playing with scientific language depends upon first controlling it. Science fiction writers like Tiptree, Asimov and Clarke were after all first scientists and then writers; they did not learn their science through science fiction.

Organizing the World: Processes

Classifying Processes

Scientists do not simply organize things, as in the examples studied above. They are also concerned with processes (what is going on). Heading *et al.*, 1967, p. 114

177

Figure 9.8:   *Processes of Change in Geology*

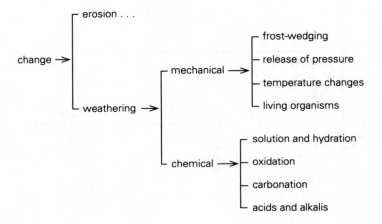

for example define geology as 'a study of the rocks and minerals of the earth's crust, and of the *changes that have occured in the rocks since the earth first came into existence* (my emphasis). Consequently a geologist classifies rocks and minerals and breaks down their composition; at the same time she is concerned with change. The concern with processes can be illustrated by considering two of the principle agents of change in geology — weathering and erosion:

> The production of rock waste by mechanical processes and chemical changes is called weathering. (p. 116)

> The destruction of a land surface by the combined effects of abrasion and removal of weathered material by transporting agents is called erosion. (p. 116)

The text then goes on to break weathering down into types according to the agents involved (three dots after a term in diagrams means that the textbook provides further sub-classification not included in the diagram). See Figure 9.8. All of the terms in this diagram refer to processes rather than things. **Carbonation** for example is the technical term for the following process (Heading *et al.*, 1967, p. 119):

> As rain falls through the atmosphere, small amounts of carbon dioxide dissolve in it. If this water lies on swampy ground where decay is occurring, it will dissolve additional carbon dioxide.

> As water containing even small amounts of carbon dioxide soaks into the ground, it immediately begins reacting with carbonate minerals. One of the commonest of these is calcium carbonate or limestone. Limestone in contact with ground water for a long time slowly reacts with it to form a solution containing calcium ions and bicarbonate ions.

. . .

The slow 'dissolving' of limestone produces sink-holes, caverns, and underground streams (figure 7-F) . . .

[the three dots in text refers to elided material]

Commonly, during the explanation of processes, technical 'things' are intro-duced — for example the **sink-holes, caverns** and **underground streams** referred to above. This means that for a geologist, part of the meaning of the term 'sink-hole' relates to the way sink-holes are formed, which relates closely to the meaning involved in classifying sink-holes and decomposing their structure. As a result, one criterion that can be used for classifying things has to do with the way in which they were formed. Geologists' classification of rocks is based on this principle. See Figure 9.9. Note that this taxonomy gives examples of each type of rock and provides some annotation of the processes involved. Teachers could usefully guide their students into some discussion of the the value of such annotations, drawing them into talk not only about where they are placed, but also about the kinds of advantages to building meaning that these offer. Such discussion can be put to good use in the students' own writing, where the use of diagrams with annotations may prove an essential part of some of the texts they need to write.

Decomposing Processes

Because carbonation is a term referring to a process, its definition is much longer than is typical when things are being defined. A number of essential steps are covered:

a. rain falls
b. carbon dioxide dissolves in rain
c. water soaks into ground
d. carbon dioxide reacts with carbonate minerals
e. sink-holes, caverns etc. form as limestone dissolves

These step-by-step definitions for processes will be referred to as explanations. The steps form an implication sequence: that is to say, if step a, then step b; if steps a and b, then step c; and so on.

Some of the clearest examples of explanations occur in meteorology. The following text breaks down the process leading to sea breezes (Messel *et al.*, 1964; Chapter 7, p. 2):

Sea breezes Sea breezes begin during the afternoons of hot days when the air over the ground becomes heated. Radiant energy from the sun is absorbed by the ground and this energy is converted into heat energy which raises the temperature of the rocks and soil. Thus the air in contact with the ground is heated, and tends to rise. At the same time, the radiant energy from the sun which is received by the sea is partly

Figure 9.9: A Geologist's Classification of Rocks

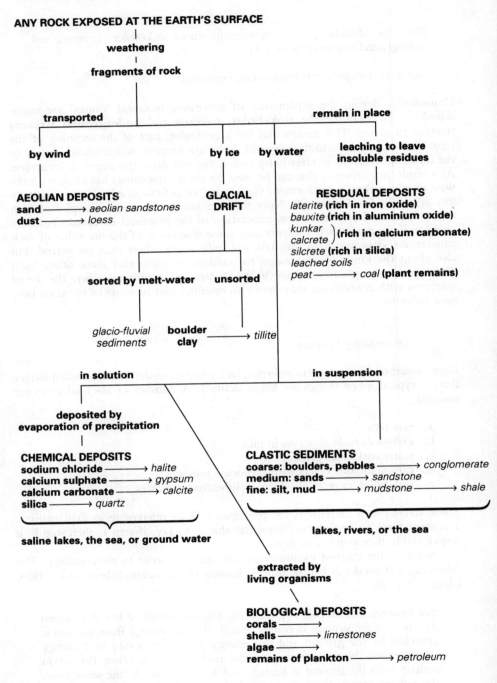

Source: Heading *et al.* (1967) *Science for Secondary Schools*, p. 308.

Figure 9.10   A Sea Breeze

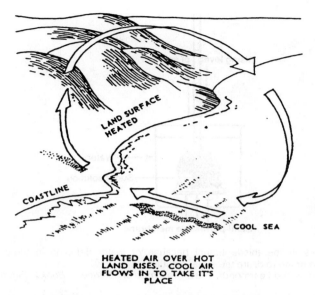

HEATED AIR OVER HOT
LAND RISES. COOL AIR
FLOWS IN TO TAKE IT'S
PLACE

Source: Messel *et al.* (1964) *Science for High School Students*

used up in converting the water into water vapour. The temperature of the sea surface does not rise as much as that of the land, because water requires more heat than other substances to produce the same rise in temperature. Thus the air above the sea is cooler than the air over the land. The result is that the heated air above the land rises, as the cooler air from the sea flows in to take its place. See Figure 7.2.

Diagrams are commonly used to represent processes such as these. See Figure 9.10.

Processes lend themselves to exemplification and experimentation. In the following passage the steps composing diffusion are explained by giving two examples, then defined (Heffernan and Learmonth, 1981, p. 156):

> If a small amount of perfume is released in the corner of a closed room, it slowly spreads so that it can be smelled in all parts of the room. If a crystal of potassium permanganate is dropped into the beaker of water, it dissolves and the purple colour slowly spreads through the beaker. Both of these cases are examples of **diffusion**. Whenever a substance is in a high concentration (Chapter 5), it will move, if it can, to a place where it is in a lower concentration.
> example 1:   perfume spreading in room
> example 2:   potassium permanganate dissolving in water
> definition: a substance in a high concentration, moving, if it can, to a place where it is in lower concentration

Figure 9.11   Osmosis

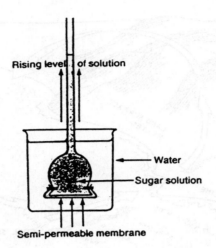

The water rises in the thistle funnel because of water diffusing in through the semi-permeable membrane to dilute the sugar solution.
Source: Heffernan and Learmonth, 1981, *The World of Science — Book 2*, Figure 9.16

Osmosis on the other hand is first explained in general terms; then a description of an experiment is provided to exemplify the process (Heffernan and Learmonth, 1981; p. 156):

A membrane which allows some substances to move across it and not others is said to be a **semipermeable** membrane. Water is one substance that moves across cell membranes fairly easily. The diffusion of water across a semipermeable membrane is known as **osmosis**. Mostly the water moves to make a highly concentrated solution more dilute. Normally, a substance such as sugar or starch would move by diffusion to lower its concentration. When a semi-permeable membrane stops this movement, the water that the sugar or starch is dissolved in moves instead. This can be seen when concentrated sugar solution is separated from pure water by a semipermeable membrane (Figure 9.16). The water diffuses through the membrane to dilute the sugar solution. This causes the water to rise in the tube. The sugar cannot pass out through the membrane.

Note that the description of the experiment is not coherent on its own. It in fact depends on the diagram. Without the diagram it is not clear what water is rising in what tube! See Figure 9.11.

One feature of scientific discourse that is highlighted in the diffusion and osmosis texts is the way in which technical terms accumulate information, allowing the chemist to move from one explanation to the next. The explanation of osmosis depends on that for diffusion, which in turn depends on the discussion of concentration and saturation in an earlier chapter. One technical term is used to define another. This is an important aspect of the condensing power of technical terms reviewed above.

TERM	DEFINITION
diffusion	the process whereby a substance in *high concentration* moves to a place of *low concentration*
osmosis	the *diffusion* of water across a semi-permeable membrane

Experiments

Scientific Method

So far we have considered the results of scientific inquiry — the kind of information involved in a scientific reconstruction of the world. Another important consideration is the way in which this inquiry proceeds. Australian science-syllabus documents all stress the value of teaching scientific method to students, involving the ability to 'recognise patterns, generalise, formulate hypotheses, observe, experiment and establish controlled situations where possible' (NSW *Science 7–10 Syllabus* 1984, p. 5). Messel *et al.* (1964, Chapter 1, p. 13), describe the 'scientific method' as follows:

> Science advances in a definite pattern. First and foremost scientists must make observations. These observations must be careful and accurate; and as the results of more and more observations accumulate in any one field they often seem to form a maze of complicated facts which are difficult to understand.
>
> It is then, however, that the second important part of the scientific method is used. Scientists look at the results of all their observations and try to develop a theory or model of how whatever it is they are investigating might work. A theory or model which is successful at the time is one which fits all the observed results and makes them seem quite natural. This model or theory can then be used to predict the results of new observations; if these turn out to be correct, the theory is further substantiated. If, on the other hand, new observations give results contrary to the model, then it must either be discarded in favour of a new model or improved.

Experiments in School

Experiments in science classrooms are designed to illustrate the way in which scientists work. They can however be used in two ways. One way, which we might call inductive, would be to use experiments to make observations for which a theory needs to be constructed by way of explanation. The second, which might be called deductive, is to use experiments to illustrate (or perhaps 'test') existing theories.

We saw earlier an account given of an experiment which exemplifies osmosis. Messel *et al.*, 1964, Chapter 5, p. 9 use the same experiment to exemplify the process, but ask students to do it rather than simply reporting what happens: (See Figure 9.12)

Figure 9.12   Osmosis

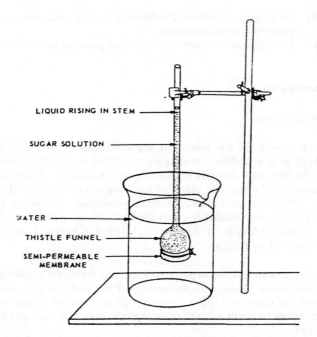

LIQUID RISING IN STEM

SUGAR SOLUTION

WATER

THISTLE FUNNEL

SEMI-PERMEABLE
MEMBRANE

Source: Messel *et al.*, 1964, *Science for High School Students*, Chapter 5, p. 11.

Osmosis occurs when a solution is separated from some of the pure solvent by a special kind of partition called a **semi-permeable membrane** . . . This simple experiment will show you an example of osmosis.

Experiment 5.20. Tie a piece of cellophane over the mouth of a thistle funnel, invert the funnel and partly fill it with a concentrated sugar solution. Make sure the string around the funnel and cellophane is tied tightly so that no leaks occur. Now place the funnel with its mouth in a beaker of water and clamp it to a stand. Set the apparatus aside and notice the level of the sugar solution in the stem of a thistle funnel from time to time during the day and also next day. See Figure 5.10 for details.

The text then points out that the water in the funnel should rise and that tasting will show that the sugar is still in the funnel, not the beaker. The explanation for this can only be that water particles have diffused through the cellophane into the funnel (osmosis) but that sugar particles have not diffused into the beaker.

The sheer volume of information prescribed for students to digest in secondary school and the high cost of much of the relevant technology means that experiments are mainly used to exemplify scientific understandings. Teachers have to be selective about which areas they choose to approach inductively, because using experiments to build up observations as the basis for constructing a theory can take a long time. The strong emphasis on processes in current Australian Science syllabi puts teachers in a difficult position. The NSW *Science 7–10 Syllabus* (1984) advises as follows:

> The proportion of time spent on practical activities will vary with the content area. However, practical work including classroom, field and laboratory activities, should form the basis for the majority of the science experiences of students. (NSW, 1984, p. 6)

The implication of this kind of advice is that individual students recapitulate the history of scientific discovery in their learning, something no professional scientist would ever attempt to achieve. What seems to be going on here is a kind of denial of the use of scientific language to accumulate and document the results of previous research in such a way that it can be taken as the starting point for new investigations. Learning scientific method is obviously important; but it is not necessary for every student to rediscover every wheel.

Cull and Comino's progressive *Science for Living* 1987 challenges the use of experiments as exemplification by setting out activities which can only be approached inductively. Consider the following experiment designed to provide observations explaining why granite crystals are large than basalt ones.

Activity 9.1 Why are the crystals in granite larger than crystals in basalt?

In order to suggest an answer to this question we shall experiment with bluestone crystals.

Procedure
* Add about a teaspoon of powdered copper sulphate to a 20 mm test tube.
* Three-quarters fill the test tube with hot water. Fit a stopper and shake to dissolve as much powder as possible.
* Allow any undissolved powder to settle and decant the clear solution into a petri dish.
* Watch the solution closely. Use a hand lens if necessary. (Cull and Comino, 1987, p. 121)

Two pictures follow, one of the solution after a few minutes, the other after a few hours. Then there are questions:

> 1 How long was it before the crystals started to appear? Were they large or small?
> 2 Did you notice anything about the shapes of the crystals?
> 3 Did any larger crystals appear? When?
> 4 Explain why granite has larger crystals than basalt?

Answers to these questions are not provided. Some indirect help is given in text preceding the experiment:

Rocks formed beneath the surface

Sometimes the magma pushes up into the crust but doesn't quite make the surface (see Figure 9.8).

Because it is now close to the surface it cools fairly quickly and turns into rock. This rock becomes part of the Earth's surface when the material above it is worn away.

For the same reason, cores of extinct volcanoes are sometimes exposed, as in Figure 9.10.

Figure 9.11 shows the whole crust pushed up without cracking. There is still a thick crust above the magma. Under these conditions the magma takes a long time to cool, and it could take a million years for it to solidify into rock. Granite is one of the rocks formed in this way.

But nowhere does that text explicitly explain that basalt has smaller crystals because it cooled more quickly than granite. The point is that it is much quicker to write a sentence or two to explain the difference in crystals than to work through an experiment and surrounding text implying this explanation (some dozens of sentences and several hours of experimenting). What worries science teachers is that if you just tell students things, they won't learn them — involvement in a process is felt to lead to 'real understanding'. The price that must be paid on the other hand for working inductively is that much less science can be taught.

Some kind of balance must be struck. The present trend is to emphasize inductive processes in primary and junior-secondary school, which puts tremendous pressure on the upper secondary to shift radically away from process and experimentation in order to make up the lost ground. In sorting this out it needs to be kept in mind that scientific language has evolved so that it can accumulate information making it unnecessary to repeat the same research from one generation to the next. Students can be taught to access these genres, beginning in infants school (see Rothery, 1986, on report writing in Year 2). The nature of these genres is explored in the next section.

Scientific Genres

Storing Facts

So far we have looked at writing in science from the point of view of field (or less technically its content or subject matter) — the way in which scientific knowledge is constructed. This information, as we have seen, can be formulated in various ways: in diagrams or in text, and if in text then in different genres. The genres are adapted to the different aspects of science that are being written about.

A necessary part of becoming a proficient science student is learning to read and write the various genres particular to science fields, and for that reason teachers need to be careful in thinking about the various genres they want their students to learn. As noted earlier, much contemporary advice to teachers does not in fact do justice to the actual genres of science. And genres more appropriate to other fields than science are recommended to be taught. This leaves many students rather uncertain about what is expected, which has very serious ramifications for individuals during public examinations, and equally serious ramifications for a 'banana republic' in which science and technology are expected to play an increasingly major role.

Reports
The major genre is science textbooks is what is technically called a *report*. The textbooks themselves are in fact large reports, broken down by headings and sub-headings into sections. The main function of these reports is to organize information about things, typically by classifying them or decomposing them. A number of examples were given above. Here is a further example each of reports concerned with classification and composition:

● Report — classifying

> Plants are divided into a number of groups (Figure 10.2). The algae, fungi and lichens are similar in that they have no real stems, roots or leaves. The **algae** are simple plants that contain chlorophyll and live in water. They vary in size from the green 'scum' growing on the walls of fishtanks to the very brown seaweed called kelp. The **fungi** are different from algae in that they contain no chlorophyll. They are able to feed on other plants and animals. Fungi vary in size from the tiny yeasts used in brewing and baking, to large mushrooms and toadstools found in fields and forests. Mould on bread and the disease 'tinea' or athlete's foot are due to fungi. The **lichens** are some of the hardiest of plants. They are a partnership of fungus and algae living together to help each other. Lichens can be found clinging to rocks and posts in most parts of the world. (Heffernan and Learmonth, 1981, p. 169)

This report defines and classifies algae, fungi and lichens, exemplifies each and gives brief descriptions of their size and habitats. The following report functions similarly, except that the terms defined are organized into parts and wholes rather than classes and sub-classes:

● Report — decomposing

> All ecosystems have certain features. Figure 14.4 shows that they all have five parts.
> 1. The Sun gives light energy to the system. No ecosystem can exist without a supply of the Sun's energy.
> 2. The green plants use this light energy, with other chemicals (water, carbon dioxide and soil minerals) to make compounds needed by each plant to live and grow. (See Chapter 10). Since they produce their own needs (except energy), green plants are called **producers**.
> 3. Animals eat these plants directly (or other animals which may have eaten plants) and are called **consumers**. They are called consumers, because they do not make their own food.

4. There are also **decomposers** present in all ecosystems. As they feed, they break down dead plants leaves, animal remains and wastes. They break them down into minerals, as well as getting their own needs.
5. These breakdown products are thus put back into the soil and can be reused. These minerals, as well as water and some of the gases in air, are called **nutrients**. (Heffernan and Learmonth 1981, pp. 241–2)

Reports typically begin with some kind of general organizing statement:

Plants are divided into a number of groups (Figure 10.2).

All ecosystems have certain features. Figure 14.1 shows that they all have five parts.

Where these are oriented to classification and composition the rest of the report is organized around the relevant sub-classes and parts. Reports have a number of distinctive linguistic features — principally:

a. generic participants (e.g., **Plants**, **ecosystems**, **animals**)
b. timeless verbs in simple present tense (e.g., **shows**, **have**, **are divided**)
c. a large percentage of being and having clauses (e.g., **have**, **are**, **are called**)

In primary and junior-secondary school most students write reports which focus on specific sub-classes, rarely focusing on taxonomies as a whole. These reports begin by situating the sub-class in the relevant taxonomy (e.g., **Dolphins are sea mammals**) and go on to describe appearance and (in the case of living things) behaviour. In the following example from a Year 6 using a genre based approach to literacy development (see Rothery, 1986) the second paragraph focuses on composition:

Dolphins are sea mammals. They have to breathe air or they will die. They are members of the Delphinidae family.

Dolphins have a smooth bare skin. Only baby dolphins are born with a few bristly hairs on their snouts. These hairs soon fall out. They have a long tail and the fin on the top of their backs keeps the dolphins from rolling over. The female dolphins have a thick layer of fat under their skins to keep them warm when they dive very deep. The dolphins' front fins are called flippers. They use them to turn left and right. Dolphins grow from 2 to 3 metres long and weigh up to 75 kilograms.

Dolphins hunt together in a group. A group of dolphins is called a pod. They eat fish, shrimp and small squid. They live in salt water oceans. Dolphins can hold their breath for six minutes.

When dolphins hear or see a ship close by, they go near it and follow it for many kilometres. Dolphins can leap out of the water and do somersaults. Sometimes they invent their own tricks and stunts after watching other dolphins perform.

Dolphins are very friendly to people and they have never harmed anyone. They are very playful animals.

Alongside classifying and decomposing, reports in science may also be descriptive, listing properties or reviewing habits/functions/uses. Two examples of descriptive reports are given below:

- Report — describing functions

   The underground part of plant is its root system. The roots have the jobs of
   a anchorage — holding the plant in the soil
   b absorption — taking in water and minerals from the soil
   c conduction — carrying water and minerals to the stem, and sap (food) from the stem back to the roots
   d storage — storing unused food for hard times, such as drought or cold winters. (Heffernan and Learmonth, 1981, p. 183)

- Report — listing properties

   This is a helpful place to make a list of the properties of living things. Living things show —

   * movement;
   * responsiveness;
   * assimilation;
   * growth with development;
   * reproduction.

   If you examine an object, and it shows all of these properties, then the object is living. (Messel *et al.*, 1964; Chapter 8, p. 2.)

This text illustrates a relatively uncommon method of definition which was not considered above — definition by accumulated properties. Like all the report genres so far examined, it is a useful one for students to be taught, and textbooks generally provide a valuable source of possible models. Teachers should be aware however, particularly at the primary level, that many texts offer poor models, and that they will need to devise models of their own for teaching purposes. Because of its fragmented process focused organization for example, Cull and Comino (1987) is a much poorer source of models of scientific writing than the more traditional Messel *et al.* (1964).

In a major project developed by the Victorian Catholic Education Office, McNamara and her colleagues worked with a number of secondary teachers in several subjects, including science. McNamara (1989) discusses the importance of models and structured questions when introducing students to report writing. She describes the work of one science teacher who wrote a sample genre showing the kind of writing he wanted and then provided students with the following questions to help structure their text:

1. How does the dung beetle use animal waste?
2. Is the dung beetle a consumer, producer or decomposer? Explain.
3. What are the limitations of the Australian dung beetle?
4. What problems developed as a result of the introduction of hooved animals?
5. Why do you think so many species were originally brought out from Africa?
6. Why were experiments carried out before releasing the beetles?
7. Explain the success of the project.

The students used the model and questions as a guide, as well as drawing on information in a video they had seen. With modelling of this kind the writing was as a whole the best the class has produced.

In addition it is important to contrast questions designed to scaffold a report from those most commonly used in science, where short answers are required to test knowledge. The philosophy behind short-answer testing appears to be that science is about learning technical terms. As we have seen, science does indeed make use of technical language; but there is far more to science than that. The relationships between terms and the criteria forming the basis of the classification and composition are essential understandings. Short answers do not provide students with opportunities to explore or display relationships of this kind. The fact that most students make very little use of report writing to integrate their understanding of classification and composition in a holistic way critically debases their socialization into scientific understandings of our world.

The following page from a Year 8 geography student's workbook illustrates the fragmented nature of much short answer writing. Each entry has been graded by the teacher. The only thing that unifies the text is its focus on technical terms associated with deserts.

Definitions

ERG — a unit of energy, amount of work done by a force of one dyne acting for a distance of one centimetre

EXFOLIATION — Process of scaling or peeling. The bark, skin or bone that is coming off in scales or peeling in layers.

NOCTURNAL — referring to animals who sleep during the day and come out at night to hunt or find food etc.

EPHEMERAL — short lifed, non woody. Fleeting, transitory, of short duration, soon passing away.

ARTESIAN — 'Artesian well' — one which is bored into a water-bearing stratum, from which a constant supply of water rises through a narrow pipe

BOLSON — these are desert basins into which desert streams drain.

SUCCULENT — juicy, full of juice or sap.

XEROPHYTE — a type of plant which has adapted to dry hot conditions

DROUGHT — exclusive dryness; spell of dry weather lasting enough for the land to become parched.

Explanations

As noted above, textbooks are basically large reports made up of a series of smaller ones. When the smaller reports focus on processes, either to classify them or to use them as criteria for classifying things, another genre, the explanation, is used. This genre differs from reports in two main ways:

a. it has a higher percentage of action verbs
b. the actions are organized in a logical sequence

Here are two further examples:

Explanation

We saw in Chapter 6 how water under increase of pressure must be raised to a temperature higher than 100°C before it will boil. In the same way rocks and material of the earth must be raised to a higher temperature than normal in order to become molten within the earth — because they are under high pressure. In parts of the earth beneath the crust, however, heat accumlates to such an extent that it does cause local melting of the rocks to form a molten mass called **magma**. This molten material is under such enormous pressure that some of it is forced into any cracks and crevices that might form in the upper solid crust of the earth, and in surrounding solid rock. Some of this molten rock can actually cool and solidify without reaching the earth's surface; in other cases molten material is pushed right through the earth's surface and forms a volcano. When molten material is forced out on to the earth's surface it is called **lava**. (Messel *et al.*, 1964, Chapter 12, p. 1.)

Explanation

As boulders and pebbles are tumbled along beds of rivers, they become rounded and smooth. Wave action along coastlines also rounds and smooth large rock fragments eroded from cliffs. Thus we find accumulations of water-worn stones close to the shores of rocky coastlines, or in beds of fast-flowing streams.

In time, finer sediment may fill the spaces between the pebbles, and become hardened by loss of water and cementation. A rock in which are embedded water-worn pebbles and boulders is called a conglomerate (figure 17-B). The water-worn stones may themselves be pieces of sedimentary rock, or pieces of granite, or possibly a mixture of rocks of more than one kind. (Heading *et al.*, 1967, pp. 291–2)

Like reports, explanations have generic (**boulders**, **pebbles**, **river** etc.) rather than specific participants and make use of timeless verbs (**become**, **rounds**, **smooth** etc.).

For many students explanations are the main source of extended writing. Here is a largely copied student text from Year 7 geography:

CONVECTIONAL RAINFALL

Air in contact with a warm surface will become heated & expand causing it to rise. As it rises it cools. When it reaches DEW POINT
(the temperature where the air is 100 per cent saturated with water) the
air CONDENSES & convectional clouds form. Thunder storms will
often result.

Exemplifying and Testing Facts

Alongside learning facts, science students as we have seen are expected to participate in experiments, which for the most part are designed to exemplify selected
facts. Accordingly, science textbooks include large numbers of procedural texts
which function as recipes for these activities:

- *Experiment — procedural*

  The following experiment shows that leaves have an important function
  in the loss of water from the plant.

  Experiment 11.3

  Cut a leafy shoot of a plant under water[2] and put the stem in water
  contained in a test-tube. Cover the water with oil to prevent evaporation
  from the surface. Support the test-tube with a bell jar (Figure 11-D). The
  ground base of the bell-jar should be coated with petroleum jelly to
  ensure a proper seal. Set up, as a control, another apparatus similar in all
  respects, except that the leaves are removed from the shoot used.

  [2]Cutting the stem under water prevents air bubbles from blocking the
  cut ends of the conducting vessels.

  After several hours, note the levels of water in the test-tubes and examine the insides of the bell-jars for droplets of condensed water.

  The results of this experiment suggest that water is lost from the plant
  by evaporation from the leaves. (Heading *et al.*, 1967, pp. 190–1)

One of the most distinctive features about this genre is the use of imperatives to
direct the student activity (e.g., **note the levels of water**, **examine the insides
of the bell-jars**, etc.).

Unlike some reports and explanations, experiments have a very clear staging
structure: Aim — Method — Results — Conclusion. Like all generic structures
this structure is functional. The Aim shows the relation of the experiment to the
scientific knowledge being constructed. The Method section provides explicit
instructions so that the experiment can be replicated. The Results stage provides
for comparability across replications. And the Conclusion relates the results to the
purpose of doing the experiment. The structure of the experiment genre
symbolizes the scientific method discussed above and has evolved to enable
scientists to document their research. [Chapter 3 contains a text from Newton
which provides a useful example of an earlier stage in this genre]

The experiment noted above clearly realizes this structure:

Aim   —   paragraph 1 (**The following experiment** . . .)
Method   —   paragraph 2 (**Cut a leafy shoot** . . .)
Results   —   paragraph 3 (**After several hours** . . .)
Conclusion —   paragraph 4 (**The results** . . .)

Less commonly an account is given of an experiment, in place of the recipe style just considered:

- *Experiment — recount*

Scientists have devised the following experiment (see Figure 11-F) to test a theory on how water can move upward in plants.

The stem of a leafy shoot was cut under water and the end inserted into a short piece of rubber tubing. The other end of the rubber tubing was connected to a 12-metre length of fine-bore glass tubing filled with water. The glass tubing was carefully lowered from the roof of a tall building and the open end of the tubing inserted into the bucket of boiled water[4] which had been brightly coloured with dye. Care was taken to ensure that there were no leaks in the apparatus.

[4]Boiling the water removed dissolved air which might have formed bubbles in the glass tubing.

The coloured water slowly began to rise up the bore of the glass tubing from the bucket. Finally, the water reached the tip of the stem, having travelled a distance of almost 12 metres.

What provided the force necessary to raise the water to this height? It is clear that atmospheric pressure is not sufficient to provide this force; you have learnt (Book 1, chapter 24) that even at sea-level, atmospheric pressure cannot force water to heights greater than about 10.4 metres.

As atmospheric pressure is the only force acting from below which could push water up the tubing, it appears that water is pulled up the tubing from above. In being pulled up the tubing, it replaces the water lost from the leaves by transpiration. This pull is known as transpiration pull. (Heading *et al.*, 1967, pp. 192–3)

The main difference here is that the imperatives in the Method stage are replaced by past tense verbs recounting what happened. Note that the participants, as well as the events, become specific in both the Method and Results stages. In the Aim and Conclusion sections they are generic, as in reports and explanations.

The text also makes frequent use of the passive in its Method section, in order to establish the following pattern of Themes:

The stem of a leafy shoot
the end (of the shoot)
The other end of the rubber tubing
The glass tubing
the open end of the tubing
which (boiled water)
care

Had these sentences been written in the active voice, then the scientists conducting the experiment would have been topical Theme in every clause. But the text is not about scientists; it is about leafy shoots in a rubber tube. So passives were selected to get the point of departure for each clause oriented to the organization of the Method stage.

Patterns such as these are crucial to effective write-ups of experiments. Yet, as has already been suggested, much curriculum advice to teachers, and consequently much teaching practice, tend not only to play down the significance of such patterns, but even to encourage children to write inappropriate ones. It is for example troubling to read the following advice to teachers:

8.2 Initially personal and informal forms of reporting are more appropriate than the classic report style. As students gain skills and confidence, the formal report format and the use of third person and passive voice can be developed. (NSW *Writing K-12 Syllabus*, 1987, p. 126)

Advice such as this is based on two unsubstantiated assumptions. This first, taken from the work of Britton *et al.* (1975) is that transactional writing grows out of the expressive. Newkirk (1984) has shown that pre-school children quite commonly write a variety of transactional texts. Factual genres do not grow out of narrative, but have their own roots in spoken and written language whose function is to explore the world (see Halliday, 1975; Painter, 1984; 1985 for explorations of this kind).

The second relates to a set of rather patronizing and romantic set of attitudes to children that are discussed in Martin (1985) under the heading 'childism'. These involve among other things the idea that science is too hard for children (especially girls, who are better suited to narrative) and that the language of science (and thus the science) has to be watered down and made personal, expressive or imaginative for children to understand it. There are now in fact a number of research studies (Rothery, 1986; Christie, 1986; 1987; 1988; Collerson, 1984) which have demonstrated that children can be introduced to factual writing from the beginning of school and that the main factor which has made it appear difficult in the past is simply that effective contexts for teaching writing have never been properly developed.

McNamara (1989) has also worked with science teachers on writing up experiments. One teacher produced a scaffolding for writing up science experiments shown in Table 9.1. Scaffolding such as this, which includes a description of the function of each stage in the genre along with some consideration of the grammar used to realize each stage is an extremely useful and powerful adjunct to the use of model texts and questions (as described for report writing above). In practice, every genre that science teachers expect their students to write needs to be deconstructed in this way and taught explicitly to students if they are really to be expected to write science.

Table 9.1: How to Write Up Science Experiments

HEADING	WHAT TO WRITE	HOW TO WRITE IT
AIM	What do you think we were trying to find out in this experiment?	Write a short, single sentence statement beginning with the word 'To . . . . . .'
METHOD	Describe in your own words exactly what you did.	1. Write numbered statements. 2. Use the word 'we' instead of 'I'. 3. Use past tense — 'was' and 'were', etc.
RESULTS	Include a table of results.	
CONCLUSION	What did you discover in this experiment?	Write a few short sentences explaining what you found.

Morris and Stewart-Dore refer to scaffolding of this kind as a *graphic outline* (1984, pp. 57–84). Their use in science lessons is again outlined in their appendix. They point out that from the point of view of reading, these outlines allow students to 'survey a text efficiently' and are very useful in summarizing, an important study skill (1984, pp. 83–4).

McNamara (1989) discusses the following Year 10 text, recounting an experiment which 'tests the ability of liquids to pass through a substance which is acting as a "stand-in" for the wall of a cell.' Note that the writing follows closely the staging of experiments as they are modelled for students in science textbooks. The passive is not used to re-organize Themes in the Method stage.

Science
Passing Through Walls Experiment

Aim. To observe how a cell membrane works. (in a similar way).
Method. In this experiment we placed two thin plastic tubes (which were acting as cell membranes) in two (separate) beakers. In these tubes were two chemicals acting as the cell's nucleus (the other insides of the cell). These two chemicals were Conga Red and Potassium Permanganate. We then filled the beakers with water and placed the tubes containing chemicals in the observations. After a couple of days we observed the changes. The potassium permanganate was starting to leak but no other changes had occured in the beaker with the tube containing Conga Red.

[diagrams omitted]

Conclusion. Our results showed that the Potassium Permanganate leaked through the tube but the Conga Red did not.
    This means that the molecules of the Potassium Permanganate are small, so they are able to fit or leak through the tube (cell membrane). Then of course this means that the molecules of Conga Red are large and cannot fit or leak through the cell membrane (tube).
    Conga Red can act as a cell nucleus unlike Potassium Permanganate.

Other Genres

Reports, explanations and experiments are far and away the most common genres used to introduce students to scientific facts and research methodology. Two other significant genres are less common: biography and exposition.

Biography

Biography is used to review the history of science. It may focus on the work of one individual; or it may cover the work of several scientists building on each other's work in a particular line of inquiry.

The following text reviews the work of cell biologists:

Biography

Although many people probably saw cells earlier, an Englishman Robert Hooke (1635–1703) was the first to realise their importance. In 1665 he was looking at a thin slice of cork under a microscope. He noticed that it was made of tiny holes marked off by walls. It reminded him of honeycomb. So he called the little holes cells from a Latin word meaning 'a small room'.

A few years later, a Dutch janitor Anton Van Leewenhoek (1632–1723) used lenses that he had made to look at many different things. In 1675 he found living things in water from a ditch. They were much too small to see with the naked eye. We now call these tiny organisms protozoa. In 1680 he found that yeasts (fungi) were even smaller than protozoa. And in 1783 Van Leewenhoek observed still smaller living things, which we now call bacteria.

After more than a century of viewing with microscopes, the idea that cells were to be found in all living things was accepted. In 1839 Theodore Schwann (1810–1882), who had studied animal cells, and Matthais Schleiden (1804–1881), who had studied plants, proposed that all living things were made of one or more cells. It was also observed that large living things, made of huge numbers of cells, began life as a single cell. This single cell could divide into two cells, the two divided and became four, and so on until the final number of cells needed were formed. In 1860, the German biologist Rudolf Virchow showed that 'all cells arise from other cells'. In 1861 Louis Pasteur (1822–95) once and for all showed that living things could only come from other living things. (Heffernan and Learmonth, 1982, p. 164)

Exposition

Exposition is a genre which is used to present arguments in favour of a position that needs to be argued for. This genre is quite rare in science textbooks. The reason for this seems to be that textbooks present scientific knowledge as certainty. Contentious interpretations are either dressed up as if uncontentious or omitted.

Exposition is used in the following text however to review the arguments in favour of reinterpreting dinosaurs as warm-blooded animals.

Exposition

Reptiles are cold blooded animals. This term (Book One) means that their body temperature depends on their surroundings. 'Warm-blooded' animals use a lot of food to keep their bodies at a certain temperature. Until recently, dinosaurs were assumed to be cold-blooded too, like other reptiles. Recent studies have put some doubt into this belief. The bone structure of dinosaurs is close to the structure of warm-blooded animals. Fossils of dinosaurs in areas which would have been very cold when they were alive have also been found. Warm-blooded animals survive much better in colder areas, because their warm bodies are able to move more quickly in the cold climate. Thirdly, warm-blooded predators need to eat more than cold-blooded predators, because they need food to keep their bodies warm. The number of fossils of other animals which some dinosaurs would have eaten, found with their bones, is larger than you would expect for cold-blooded animals. (Heffernan and Learmonth, 1981, p. 232)

The Thesis of this exposition is that dinosaurs were in fact warm-blooded. Three arguments are used to support this proposition:

Argument 1 — bone structure like warm-blooded animals
Argument 2 — fossils found in very cold areas
Argument 3 — large numbers of bones of other devoured dinosaurs found with fossils

Narrative (Imagination)

Narrative writing is not found in science textbooks; but the recent NSW *Writing K-12 Syllabus* contains the following list of recommended forms of writing in science, putting narrative at the top of the list:

* stories in Science and Agriculture, especially science fiction
* personal reports and descriptions of observations
* plays, poems and cartoons
* formal reports on laboratory work and excursions
* definitions
* essays and extended answers
* explanation of own understanding
* designing science puzzles and board games
* analysis of data
* notemaking and lecture record keeping

* keeping a diary or learning journal
* graphic presentation
* constructing an interview questionnaire
* designing advertisements or captions for cartoons or photographs

(NSW, Department of Education, (1987, p. 125)

The syllabus further recommends that students use 'imaginative writing; for example, I am Joe's heart, or the biography of a molecule of carbohydrate through the digestive tract.' (p. 125) As an example of personal imaginative writing of this kind the syllabus includes the following narrative, which appears to be an excerpt from a longer text:

Journey to the Brain

O.K. boys, on the count of three. One, two, three . . . . . .
Oh no, the ear flaps have caught my vibrations. NOT AGAIN. I shoot through the auditory canal at very high speeds, going along bumps and ridges, through every nook and cranny. Then SMASH. I hit the ear drum, jarring my whole body and making my head spin like a merry go-round. Then without any rest, I collide with three other bones, all pushing into one another. On again off again and here I go. This time to a roundabout which I must add has some greenish-grey, gooey fluid. Oh gross. (NSW, Department of Education, 1987, p. 127)

It is instructive to compare this text with a report and explanation from Cull and Comino 1987, pp. 98–9:

How does the ear work?

The outer ear

This is the part we can see. It is shaped to guide the sound into the ear and along the **auditory canal**. This canal is lined with hairs and produces wax to keep out insects and protect the ear from infection.

The middle ear

This is a tiny air-filled chamber about the size of an aspirin tablet. It is separated from the outer ear by the **eardrum**, a thin membrane.

An air passage from the throat, the **Eustachian tube**, allows air to enter the middle ear to equalise pressure on both sides of the eardrum.

In the chamber are three tiny bones that conduct the sound across the middle ear.

The inner ear

Unlike the air-filled middle ear, the inner ear is filled with fluid.

The **semi-circular canals** control the body's balance.

The **cochlea** contains the nerve that sends the impulse to the brain.

How is sound heard?

Sound waves pass along the outer ear to the eardrum. The waves strike the eardrum, causing it to vibrate. These vibrations start the first bone of the middle ear vibrating. The first bone sets off the second, which in turn sets off the third.

The bones are connected like levers.

The third bone triggers vibrations in the liquid of the inner ear. These vibrations set up electrical pulses in the cochlea. The electrical pulses travel along the auditory nerve to the brain, which interprets them as sound.

The text is accompanied by a cross-sectional diagram of the ear with the outer ear, eardrum, middle ear, Eustachian tube, small bones, inner ear, cochlea (which is mislabelled) and auditory nerve labelled.

By comparing these two texts, we can determine just how much science is included in the narrative passage. First, it contains four technical terms referring to parts of the ear: ear flaps (or more technically **pinna**), auditory canal, ear drum (or more technically **tympanum**) and three bones (more technically the ear **ossicles**; more particularly the **hammer**, **anvil** and **stirrup**). None of these terms is defined, although their relative position in the ear can be deduced from the time-line of the story. The function of the ear flaps (**catching my vibrations**) and possibly of the small bones (**pushing into one another**) is mentioned, and the fluid contents of the inner ear (**the roundabout**) described.

The real test of this content would be to see how much of Messel *et al.*'s ear diagram could be reconstructed from the narrative, and how much from the Cull and Comino text: (See Figure 9.13). A further test would be to ask how much of Cull and Comino's explanation of how sound is heard can be reconstructed from the narrative.

White and Welford, 1987 make the further point that a strong personal response, empathizing with the subject matter actually leads writers to make inaccurate observations in a task requiring students to compare and contrast two insects: 'The point to be stressed is that once committed to the narrating of impressions and personal fantasies, the possibility of introducing anything so bald as figures or measurements recedes even further. In this respect the narrative is a restrictive mode for a task in which precise descriptions are required.'

It is only through careful comparisons of narrative and scientific discourse that the real value of using imaginative narrative to learn science can be determined. It is already clear that narrative itself is a tremendously inefficient way of exploring the ways in which science interprets the world and positively distracts students from building up scientific understandings. It is not the function of narrative to classify, decompose, measure and explain. For this, the genres science has itself evolved, are naturally far more appropriate.

Figure 9.13: The Ear of Man

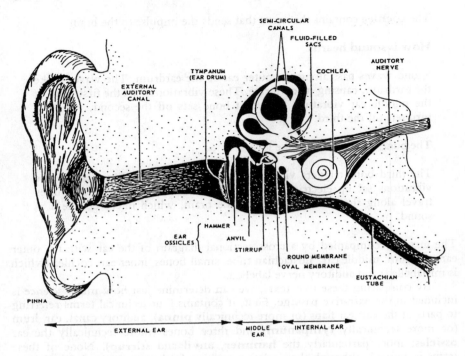

Source: Messel *et al.* (1964), *Science for High School Students*, Figure 29.14

As a final note of caution, it is very damaging to assume that all students are effective writers of narrative, in spite of the fact that recent syllabus documents take as given that personal expressive imaginative writing comes naturally to all students. Many writers, and in our culture this is partly conditioned by gender, are much more comfortable with factual than expressive genres. Syllabus documents emphasizing the role of personal writing in leading children into science are doing a great disservice to students like Ben who are more interested in the nature than nurture of things.

Science Literacy

We have pointed out that in science language is a fundamental tool. It is used to classify, decompose and explain, and to recount the investigations that form the basis of a scientific world view. It follows that to be illiterate in science is to be denied access to a crucial aspect of its technology. Most significantly it follows that science cannot be understood 'in your own words'. It has evolved a special use of language in order to interpret the world in its own, not in common sense, terms.

We have also pointed out that literacy in science has to be considered both from the point of view of field (the knowledge that is being constructed) and genre (the global patterns of text organization that package this knowledge). This distinction corresponds to Morris and Stewart-Dore's (1984) development of the structured overview to focus on field and the graphic outline to deal with genre. Most science teachers focus mainly on field rather than genre, although many do explicitly teach the structure of writing up experiments.

In spite of the focus on field very little writing is used to extend or consolidate knowledge in this area. For the most part field is explored orally; what writing there is is restricted to definitions, short answer questions, fill in the blank exercises and the like. The main function of this writing appears to be that of testing student's understanding of relatively isolated technical terms. When students are asked to write at length in projects or research reports they work on at home in evenings and on the week-end, for the most part they end up copying, with minor editorial adjustments, passages of text from research materials they have found (Wignell, 1987). Nationally (McNamara, 1989; Christie and Rothery, 1989) and internationally (Spencer, 1983; White and Welford, 1987) the picture appears to be much the same.

Recent syllabus documents reflect an increasing emphasis on doing as opposed to learning science. The Victorian Primary Science Syllabus (1981; Section 2.0) describes science as a 'practical process':

> The role of science in the primary school is to provide children with opportunities to investigate their world in an ordered way. Investigating and manipulating everyday things allows children to collect information about their environment. When this information is collated and tabulated conclusions can be drawn. Conclusions based on ordered information can be readily tested by others duplicating the same activity.

Here we have the model of the child as mini-scientist, participating in activities analogous to those undertaken by scientists centuries ago as they began to formulate their picture of the world. The effect of the advice given is to foreground doing (observing and experimenting) and background language, especially written language (reports and explanations). This means that children are not taught to access the genres science has evolved to store information which leads to a tremendous inefficiency in the science curriculum.

The NSW *Science 7–10 Syllabus* takes a more balanced view (1984, p. 2), placing equal emphasis on learning scientific method alongside scientific knowledge:

> Science can be defined as a body of collected knowledge comprising the interconnected sets of principles, laws and theories that explain the Universe. People who take this view refer only to science content — the facts, principles and laws used to describe the world around them.

> Science can also be defined as a set of processes that can be used to systematically acquire and refine information. People who take this view consider the scientific enterprise to be a set of processes for obtaining information.

For the purposes of this document the definition of Science encompasses both points of view because one point of view cannot be learned or understood without the other.

With the appearance of recent textbooks such as Cull and Comino, 1987, *Science for Living* series however, it would appear that an emphasis on science as activity is gaining momentum, particularly in primary and junior-secondary schools. Compared with Messel *et al.*, 1964, the text emphasizes activities, inductive questioning sequences, diagrams and photographs. Writing in report and explanation genres is very fragmented. This kind of textbook may be a useful adjunct to current teaching practice. But it does not provide satisfactory models of extended writing in science; nor does it provide useful reference material, since every topic is dealt with in such a partial way.

What seems to have gone wrong in the development of science textbooks over the years is that an attempt has been made to make science more accessible by downplaying science literacy. But diluting scientific discourse necessarily involves diluting the science that is taught. As we have seen, science is unthinkable without the technical language science has developed to construct its alternative world view.

To rehabilitate literacy in science teachers and students will have to work towards a much clearer grasp of the function of language as technology in building up a scientific picture of the world. Technical language has evolved in order to classify, decompose and explain. The major scientific genres — report, explanation and experiment — have evolved to structure texts which document a scientist's world view. The functionality of these genres and the technicality they contain cannot be avoided; it has to be dealt with. To deal with it teachers need an understanding of the structure of the genres and the grammar of technicality. With this knowledge they can begin to tackle the problem of science literacy along the lines of the work of Christie and Rothery, McNamara, Morris and Stewart-Dore reviewed above. Without it they will continue to focus on content without taking language into account, probably with an increasing emphasis on science activities rather than science texts. The linguistic technology is the key — not just to science literacy but to understanding and practising science itself. Ways must be devised to provide access to this technology. And the answer must not involve watering the technology down.

Chapter 10

Technicality and Abstraction: Language for the Creation of Specialized Texts*

J.R. Martin

Specialized Knowledge

At the CSIRO in Canberra, one section of the Personnel Management Manual admonishes administrators to avoid the use of the *passive tense* in their writing. As a writing consultant, I find what I suspect the author means poor advice — the passive is after all an important resource for organizing Themes in the development of an effective text. As a linguist, I find the advice simply amusing — because there is no such thing as a passive tense. There are present, past and future tenses, and active and passive voices. Presumably the would-be prescriptive grammarian who worked on that section of the manual intended the passive voice. But as the manual stands, the advice, poor though it is, does not really make sense.

In order to understand technical discourse, it is important to understand exactly what went wrong with this advice. To do this we need first to examine the nominal group **the passive tense**. In such groups the word **passive** sub-classifies **tense**. The same meaning can be re-expressed as a clause: **The passive is a kind of tense**. Technically words functioning like **passive** are called 'Classifiers' and the words they classify 'Things'. Here are some more examples (Classifiers in bold):

material process	**frying** pan
relational clause	**steel** wool
attitudinal epithet	**stone** wall
embedded clause	**red** wine
possessive pronoun	**lap** dog

Classifiers are nouns, verbs or adjectives and in English precede the noun they sub-classify. They are easily recognized because they are not gradable — that is to say to *a very possessive pronoun (c.f. a very possessive parent) or *a more

* This chapter is taken from *Writing in Schools: Reader* (1989a), School of Education, Deakin University, Geelong, Victoria, Deakin University Press, Chapter 2.3.

Figure 10.1:   English Systems for Tense and Voice

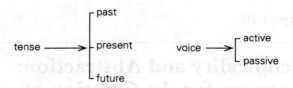

material clause (see **a more material solution**). This is because Classifiers classify; they do not describe.

When we say Classifiers classify we mean that they refer to classifications of experience known as *taxonomies*. You may be familiar with these from biology, where living things are often presented in diagrams organized into species, genus, family and so on. When someone refers to the passive tense they are implying a taxonomy like that following:

$$\text{tense} \rightarrow \left[\begin{array}{l} \text{active} \\ \text{passive} \end{array} \right.$$

But this is in fact a confusion of two taxonomies, one for voice and another for tense. Linguists would organize English as Figure 10.1.

So when I wrote that there was no such thing as a passive tense, I meant that in the field of linguistics there are no taxonomies for English in which passive is a sub-class of tense.

Knowledge of this kind is admittedly specialized knowledge, although a generation ago it was not quite so specialized as it appears today!

Common Sense

Common sense, like specialized knowledge, makes use of nominal groups (as illustrated above) and relational clauses (the 'being' clauses in the examples below) to classify experience. This is strikingly clear in the language of young children as they sort out the world (data from Clare Painter):

[3 years, 8 months]
(considering a jigsaw puzzle)
Child: There isn't a fox: and there isn't — Is a platypus an animal?

[3 years, 7 months]
(mother makes reference to her best boys)

Older brother:	We have to be your best boys cause we're your only boys.
Child:	And Daddy.
Older brother:	He's not a boy, he's a man.
Child:	He **is** a boy cause he's got a penis.

Figure 10.2: Commonsense Taxonomy of Diseases (Courtesy of Joan Rothery)

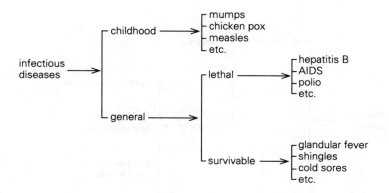

Here we see the child sorting out the world of living things, organizing animals and humans into common-sense taxonomies. Caregivers interact continually to guide the child into the common-sense organization of the adult world:

[3 years, 6 months]
Child: Mum, is grow mean move?
Mother: Well, it's not the same as move: it means get bigger, like when you blow a balloon up.

[3 years, 7 months]
Mother: Oh, we're drowned!
Child: What does drown mean?
Mother: Means we're all wet . . .
Child: Not drown; drown is go down to the bottom and be dead.

The main difference between common-sense taxonomies and specialized ones is that common-sense classification is based on what can be directly observed with the senses. Diseases for example are commonly classified according to symptoms and effects: (see Figure 10.2).

In the field of medicine on the other hand, diseases are organized according to their cause. Of the diseases noted above, cold sores, glandular fever, chicken-pox and shingles are all caused by the herpes virus, and chicken pox and shingles are in fact caused by the same virus. So the specialized medical taxonomy differs from the common-sense one: (see Figure 10.3).

Two important points about technical discourse should now be clear. The first is that technical terms cannot be dismissed as *jargon*. Technical terms organize the world in a different way than do everyday ones. Referring to a disease with respect to the simplex herpes virus is quite different to naming it cold sores. [Technical discourse can of course be used to exclude; and people are quite justified in complaining when it does so needlessly.] The second is that specialized knowledge is not just a set of technical terms. The terms imply taxonomies which

Figure 10.3: *Medical Taxonomy of Diseases*

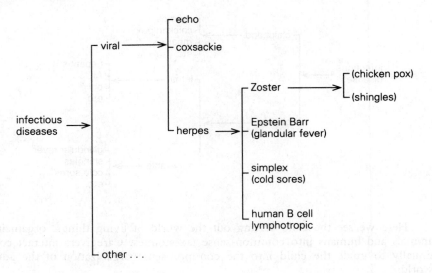

organize reality differently to common sense. Understanding technical discourse means being familiar with these specialized taxonomies and the principles which led to their construction.

Technical Writing

Technical writing is most strongly associated in our culture with the various fields of science. In broad terms science is concerned with: how the world is organized; and how it came to be that way. High-school textbooks deal with both these concerns.

The two most relevant genres will be referred to here as *report* (how the world is organized) and *explanation* (why it is organized that way). The main difference is that reports focus on things while explanations focus on processes. Science textbooks shift from report to explanation as appropriate when building up chapters.

Reports

Reports in science focus on classification and description. Consider the following section from *Our Changing World*, a geography textbook used in some Australian junior-secondary schools (generally Year 8):

Fifteen per cent of the world's land area consists of deserts. The true **hot deserts** straddle the Tropics in both hemispheres. They are found on all continents between the latitudes of approximately 15 to 10 degrees,

Figure 10.4: Taxonomy of Ecosystems.

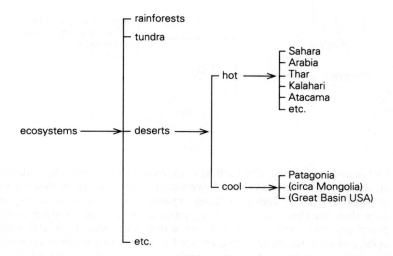

and they extend inland from the west coasts to the interiors of these continents. They are never found on east coasts in these latitudes as all east coasts receive heavy rains from either on-shore trade winds or monsoons.

Cool deserts are found further polewards in the deep interiors of large continents like Eurasia or where mountains form **rain-shadows**, which keep out rain-bearing winds that might otherwise bring wet conditions.

There are five major hot desert belts in the world (see Fig. 3.2). The largest hot desert extends from the west coast of North Africa eastwards to Egypt and the Red Sea — this is the great Sahara that covers 9 million square kilometers. The Sahara spreads eastward beyond the Red Sea into Arabia — a desert of 2.5 million square kilometers — and beyond the Persian Gulf in to Iraq and Pakistan (Thar Desert) . . . (Sale *et al.*, 1980, pp. 45–46)

This report sets forth a classification of deserts, one of the world's major ecosystems. In this text technical terms are highlighted with bold face (**hot deserts, cool deserts, rain shadows**). The taxonomy is being organized as in Figure 10.4.

So far we have only considered taxonomies which organize the world into classes and sub-classes. but reports are also commonly used to establish relations between parts and wholes. The following report examines the distinctive parts of cacti.

The conservers are those plants which store their own supplies of moisture for use during drought. These include the one thousand varieties of

Figure 10.5:   Taxonomy of Conservers: Cacti

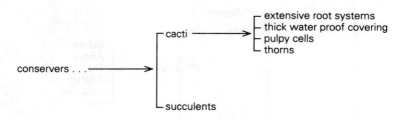

cactus and succulents. The cacti have extensive root systems spreading in all directions — sideways and downwards — to soak up as much water as possible when it does rain. They are able to swell to store water, and they then use this water over long periods of drought. A thick water-proof covering protects these desert water-tanks with their soft pulpy cells, and their leaves are often reduced to thorns to cut down on water-loss and protect the plant from animals that might otherwise eat it for its moisture (see Fig. 3.21). The giant saguaro (pronounced say-w'are-oh) of south-west USA (see Fig. 3.18) is a good example. (Sale *et al.*, 1980, p. 59)

As is the case with this report, part–whole relations are often illustrated with a labelled diagram. The relevant taxonomy is as in Figure 10.5.

Explanation

Explanations focus on processes — on how things come to be. In the geography text from which examples are taken here they tend to occur now and again in the middle of reports. Their purpose is not directly to classify phenomena but to outline an activity. The following paragraph illustrates the genre:

We saw that **leaching** was a very prominent process in all hot, wet, forest lands; in deserts, because the rainfall is so low, it hardly occurs at all. Instead, a reverse process may develop called **calcification**. Water may soak into the ground after rains and dissolve mineral salts in the usual way, but as the surface dries out, this water is drawn upwards like moisture rising through blotting paper. The salts then accumulate in the surface soil as this moisture evaporates; thus desert soils are often rich in mineral salts particularly calcium, sodium and potassium. Provided the salts are not too concentrated (and their concentration is reduced under irrigation), they contain a plentiful supply of plant foods and can there-fore be considered as fertile soils. (Sale *et al.*, 1980, p. 55)

This explanation focuses on the technical process calcification, breaking it down into steps:

Figure 10.6:   Taxonomy of Desert Landforms

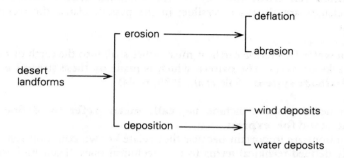

1   rain falls
2   water soaks into ground
3   water dissolves mineral salts
4   surface dries out
5   water drawn upwards
6   salts accumulate as moisture evaporates

In geography, phenomena may be classified according to the processes which gave rise to them. Desert landforms for example are organized as in Figure 10.6.

Explanations are very common when phenomena are classified in this way. Consider abrasion in the following explanation, noting once again the steps in the process:

> **Abrasion** occurs when lots of hard sand particles are carried by desert winds and are thrown with great force against all solid objects in their paths. Stones, rocky outcrops, and all natural and man-made objects are subjected to this **sand-blast** action. Within a metre or two of the ground, objects are cut, smoothed and polished by abrasion. Where winds are funnelled through gaps and valleys, this sand blast action affects all rock surfaces. Stones and rocks have their sides smoothed and rocky outcrops gradually become honeycombed with sand blasted **caves** and **windows** during successive sand storms. **Mushroom rocks**, **window rocks** and **natural arches** are produced. (see Fig. 3.5.) (Sale *et al.*, 1980, p. 51)

The Grammar of Technicality

A brief note on the grammar of technicality is perhaps in order here. First, definitions. Definitions are a special type of relational clause which in effect translate common sense into specialized knowledge. Consider the following:

> **Precipitation** *refers to* all forms of water which fall (precipitate) from the sky. (Sale *et al.*, 1980, p. 44)

Here the technical term **precipitation** is related to its definition **all forms of water which fall from the sky** by the relational process **refers to**. These defining clauses are always reversible; in the passive clause the technical term comes last:

> When water falls to the earth, it must either soak into the earth or run off in creeks or rivers. The pattern which is made by these streams *is called* the **drainage system**. (Sale *et al.*, 1980, p. 44)

Common defining verbs include **be, call, mean, refer to, define, signify, represent, stand for, express**.

Definitions are important because they relate known common-sense terms or previously defined technical terms to new technical ones. Their limitation is that they do not relate the new technical terms to each other. This as we saw earlier is the job of another type of relational clause and of nominal groups. For example when the child asked his mother 'Is a platypus an animal?' above he was not trying to define a platypus, but was trying to find out how the platypus is related to other living things.

The relevant relational clause is not reversible. The sub-class comes first and is related to its superordinate, normally by the verb **be**:

> The Amazonian rainforest is a functioning ecosystem. (Sale *et al.*, 1980, p. 24)

Clauses such as these construct classifications. Where classifications have already been established, and simply need to be referred to, nominal groups are used (e.g., **the Amazonian rainforest ecosystem** following the above):

> The heavy rainfall totals of the Tropics are related to the constant heat which occurs near the equator. In these **low latitudes**, the noonday sun is almost directly overhead throughout the year, so the sun is very effective in heating land and sea. (Sale *et al.*, 1980, p. 10)

In this text the first sentence refers to latitudes near the equator; in the following sentence these are then picked up with a Classifier Thing structure: **low latitudes**.

Part–whole relations are also contracted through relational clauses, of the possessive type, with the whole before the part:

> The biosphere is made up of many hundreds of different **ecosystems**. (Sale *et al.*, 1980, p. 12)

Once constructed possessive nominal groups can be used to refer to a whole's parts: **the Amazonian ecosystem's biome, the biome's flora, the fauna's members** etc.

In order to understand technical discourse both the definitions and the relationships among what is defined are critical. In geography, technical terms are highlighted as they are defined and commonly tested in homework, quizzes and examinations. Part–whole relationships are also commonly foregrounded in illustrations and diagrams. What tends to be backgrounded are the class/sub-class taxonomies reviewed above, although headings and subheadings often refer to global organization of this kind. These also lend themselves to the two dimensional displays commonly used by linguists (as in the figures throughout this paper) and biologists and such presentations could be made more use of across science teaching than is presently the case.

The grammar of explanations depends much more on material (action) clauses that on relational clauses. Things happen and act on other phenomena. The actions themselves stand in an if/then relationship to each other. Accordingly we might re-express the calcification text as follows (logical relationships of implication are italicized and action processes appear in bold face):

1 *If* water **soaks** into the ground
2 *then* it will **dissolve** mineral salts.
3 *If* it does, *then if* the surface **dries out**
4 *then* the water is **drawn** upwards.
5 *If* it does, *then if* the water **evaporates**
6 *then* salts **accumulate** in the surface soil.

Abstraction

On the 26 April 1986 the *Adelaide Advertiser* published reactions to Anzac Day by two students, one aged 8, the other 16. The 8 year-old began as follows:

Today I watched the Anzac parade and I saw lots of brave men and women. Most of them had medals and wore uniforms. Some drove in cars because they were too sick to walk. There were lots of countries marching apart from Australia. There were bands and thousands of people watching and clapping.

The 16 year old chose a more 'written' style:

The atmosphere at the dawn service was one of **solemnity**, as those who had first-hand **experience** of the **devastation** of war reflected on the past and remembered their friends and relatives who had lost their lives in **battle**. The gloomy atmosphere was emphasized by the dreary **drizzle**, drab attire and the long **silence**.

What is the difference between the two texts? Comparing clause Themes provides a partial answer:

Themes

8 YEAR OLD	16 YEAR OLD
Today	The atmosphere at the dawn service
and I	as those who had experience of the devastation associated with war
Most of them	The gloomy atmosphere
because they	
Some	
There	
There	

The 8 year-old uses simple 'spoken' Themes, organizing his text around people: **I, most of them, they, some** (of them). The 16 year-old uses more abstract Themes, organizing her paragraph around a description of the atmosphere at the service. In order to do this she makes use of words like **solemnity, experience, devastation, battle, drizzle** and **silence** which are referred to as nominalizations. If we remove these from the text it becomes much more 'spoken':

> People were solemn at the dawn service as those who had experienced first-hand how devastating war can be remembered their friends and relatives who had lost their lives fighting. It was drizzling, people wore drab attire and they were silent for a long time, all of which made me feel even more gloomy.

And the text had been in fact re-organized as well. It no longer focuses on *the atmosphere* as Theme. People, rather than abstractions, are the point of departure for most clauses. Comparing the 'spoken' and 'written' versions of the 16 year-old's text is a useful point of departure in interpreting the function of abstraction in writing across subject areas.

Christie and Rothery (1989) point out the connection between choice of Theme and improving a text's interpretative focus. Each of the Themes they suggest involves abstract language. Note the nominalized language in their examples (nominalizations in bold face; Theme in italics):

> *The* **desire** *to be a member of a prestige group* often influences adolescent **behaviour**.

> *Peer group* **pressure** affects young people like Debby.

> **Experimenting** *with drugs* is a dangerous thing.

Abstraction in Science

As noted above technical language enables scientists to reclassify the world. The taxonomies they establish in fact organize all phenomena as if they were things — because it is things rather then processes which lend themselves most readily to categorization as classes and sub-classes and as parts and wholes. So when processes are being classified they are nominalized and organized as things. The taxonomy cited above related to desert landforms illustrated this point; nominalizations involved included **erosion, deposition, deflation**, and **abrasion**.

Somewhat ironically then, although science does concern itself with processes, analysing them in explanations, in the end it interprets processes as things. Technical verbs are very rare, and those that exist are seldom used; for example **precipitation** is much more common than its verbal form **precipitate**:

> **Precipitation** refers to all forms of water which fall (precipitate) from the sky. (Sale *et al.*, 1980, p. 44)

Aside from facilitating classification, technical terms for processes function as a kind of short-hand. It is quicker to refer to **leaching** and **calcification** by name than to run through the processes to which each refers. In science then nominalization is used to facilitate classification. The vast majority of technical terms are nouns, and when processes are classified (e.g., **leaching** and **calcification**) or used as a classificatory principle (e.g., **deflation** and **abrasion**) nominalization is used.

Abstraction in The Humanities

Unlike science, disciplines like English and history are not very technical. They do not have as their main function reclassifying experience. And so technical terms do not present much difficulty. But abstraction in the humanities can be very challenging. Literary criticism and historical interpretation may in fact be much more heavily nominalized than scientific writing, and so no less of a problem for students to learn to read and write. For many students abstraction probably forms more of a problem than technicality, since science teachers do teach to the concepts and terms that make up scientific discourse whereas English and history teachers do not focus explicitly on nominalization as their main interpretative tool.

It should be noted however that science teachers make much more use of talk than writing to unpack technicality. In general science students write many more short single sentence definitions than reports or explanations, although they must certainly learn to read the latter where text book material is used.

The result in English, history and the humanities-oriented parts of social science is that many students continue to write as they talk. The following text was written in year 10 by a student whose writing does not sound much more mature than that of the 8 year-old quoted above.

> I think Governments are necessary because if there wasn't any there would be no law people would be killing themselves. They help keep our economic system in order for certain things. If there wasn't no Federal Government there wouldn't have been no one to fix up any problems that would have occured in the community. Same with the state Government if the SG didn't exist there would have been noone to look after the school, vandalism fighting would have occurred everyday. The local Government would be important to look after the rubbish because everyone would have diseases.

If pressed, most teachers would probably criticize this text for its spelling, punctuation and grammatical usage alongside a general comment on 'poor ideas'. In fact editing the spelling, punctuation and usage does not transform the text into a mature piece of writing in Year 10 geography:

> I think Governments are necessary because if there weren't any there wouldn't be any law: people would be killing themselves. They help keep our economic system in order for certain things. If there wasn't any Federal Government there wouldn't be anyone to fix up any problems

that occur in the community. It's the same with the State Goverment —
if they didn't exist there wouldn't be anyone to look after the schools.
Vandalism and fighting would occur everyday. The Local Government
is important to look after the rubbish, because otherwise everyone
would have diseases.

Organizing Exposition

To effect a real transformation we need to organize the 'spoken' text into exposition, highlighting its thesis, setting forth the arguments and possibly summing
up. And to do this we need abstract language — of the kind underlined in the
next version of the text below — along with appropriate linking words (i.e., **to
begin, similarly, finally**):

> I think Governments *at different levels* are necessary *for a number of reasons.*
> They make laws, without which people would be killing themselves and
> help keep our economic system in order.
>
> To begin, the Federal Government fixes up problems that occur in the
> community.
>
> Similarly, the State Government looks after schools, preventing vandal-
> ism and fighting.
>
> Finally the Local Government is important to look after rubbish; other-
> wise everyone would have diseases.
>
> *As a result of these and other factors*, Governments at several administrative
> levels are necessary.

This text still has a long way to go — as a next step its central paragraphs need
developing. But the basic scaffolding is there. In a sense what we are looking at
here is a movement from what Christie and Rothery (1989) call judgments — a
movement from personal feelings and their motivation to public positions and
their rationale.

Exposition is an important tool in interpretation in the humanities. A mature
form, taken from Simmelhaig and Spenceley, a progressive Year 10 history text,
is presented below (in the following texts nominalizations are in bold face to
highlight the abstract language used):

> Wars are costly **exercises**. They cause **death** and **destruction** and put
> resources to **non-productive uses** but they also promote **industrial
> and technological change**. This **benefit** does not mean that war is a
> good thing, but that sometimes it brings **useful developments**.
>
> The Second World War further encouraged the **restructuring** of the
> Australian economy towards a **manufacturing basis**. Between 1937
> and 1945 the value of **industrial production** almost doubled. This
> **increase** was faster than otherwise would have occured. The **momentum**
> was maintained in the post-war years and by 1954–5 the value of

manufacturing output was three times that of 1944–5. The **enlargement** of Australia's **steel-making capacity**, and of chemicals, rubber, metal goods and motor vehicles all owed something to the **demands** of war. The war had acted as a hot-house for **technological progress** and **economic change**.

The war had also revealed **inadequacies** in Australia's scientific and research **capabilities**. After the war strenuous **efforts** were made to improve these. The Australian National University was established with an **emphasis** on **research**. The government gave its **support** to the **advancement** of science in many areas, including agricultural **production**. Though it is difficult to disentangle the **effects** of war from other **influences**, it is clear that future generations not only enjoyed the **security** and peace won by their forefathers but also the **benefits** of war-time **economic expansion**. (Simmelhaig and Spenceley, 1984, p. 121)

The first paragraph articulates the Thesis — that wars promote industrial and technological change. Then two Arguments illustrating this are presented in paragraphs two and three: re-structuring towards a manufacturing basis and efforts to improve scientific and research capabilities. Both Arguments are introduced with a topic sentence and then elaborated. The exposition ends with a Reiteration, summing up the position: the benefits of war-time economic expansion.

This text is heavily nominalized, and its organization into Thesis – Argument – Reiteration depends on abstract language:

- *Thesis* — but they also promote **industrial and technological change**.
- *Argument 1* — The Second World War further encouraged **the re-structuring** of the Australian economy towards a **manufacturing basis**.
- *Argument 2* — The war had also revealed **inadequacies** in Australia's scientific and research **capabilities**.
- *Reiteration* — it is clear that future generations not only enjoyed the security and peace won by their forefathers but also the **benefits** of war-time **economic expansion**.

Abstract Reports

Reports are an important feature of writing in the humanities as well as in science. But the reports are much less technical, having the function of making generalizations about generic classes of participants. They vary in the amount of nominalization they use. The following report, which discusses three major roles of Vietnamese women soldiers, is relatively concrete:

It was mainly after 1965 when the US sent troops into the south on a massive scale, that the movement to have women join the army increased. Women became full-time members of the armed forces in the south. Many of them held leading positions. About 40 per cent of the regimental commanders of the PLAF were women. These were troops who dealt with the American mobile reserves, initiated offensive operations and attacked major US concentrations. All were volunteers who received no salary and when not in combat, helped in harvesting, building homes and schools and administering free medical care and medical training.

Women also formed a major part of regional guerilla forces, full-time fighters who operated in the region where they lived. They engaged US forces in the same area by ambushes, encircled bases and attacked posts.

Women in local self-defence units, or militia women, were not full-time soldiers but fought when their area was attacked, pinning down local forces and keeping their posts permanently encircled.

A higher percentage of women were in the local militia and regional guerilla units than in the PLAF. The local militia kept villages fortified with trenches, traps and spikes. These defences were decisive in wearing down the morale of Saigon and US troops. (Simmelhaig and Spenceley, 1984, p. 172)

The report classifies the women into three groups: PLAF, guerilla units and local militia; but it does not set up technical terms for these. It is basically concerned with describing the habits of each group.

Much more abstract reports are commonly found, both in primary and secondary sources. In history, the most abstract secondary reports function as introductions — to chapters or to collections of primary-source material. The following is taken from Allport and Allport, a progressive senior-secondary text. It discusses Masters and Servants legislation:

However, there was **uncertainty** in some quarters concerning the **regulation** of free labour. British law was applicable but colonial **regulations** were considered necessary to clarify the position. In 1828 the first Master and Servants **legislation** was enacted. This did not solve the problems of **control**; labour was still scarce and the legal system and the **policing** of its **regulations** was still at a rudimentary **stage**. By 1845 the **growth** of many working class **organizations** — **benefit** societies, trade societies, craft unions and political organizations such as the short-lived Mutual Protection Association — stimulated the **revival** of a system of **control**. All previous Masters and Servants Acts were thus superseded by the 1845 Act. The 1845 Masters and Servants Act originally aimed to provide more severe **penalties** for **breaches** of work **contract**; yet due largely to public **pressure** by the working class the bill was amended to eliminate the sections to which the workers objected. (Allport and Allport, 1980, p. 8)

Some of the most incongruent report writing in fact occurs in primary sources where various forms of bureaucratic writing are introduced. Writing in administrative contexts, with an eye to social control, is the source of much of the most heavily nominalized discourse in western culture.

The following resolutions were carried at the Australasian Conference of Employers held in the Sydney Chamber of Commerce, 10–13 September 1890.

That this Conference reaffirms the principle of '**freedom of contract**' between individual employers and employees, and asserts that any **infringement** of that principle is not only destructive to **commerce** but is also inimical to the best **interests** of the working classes.

That any **attempt** to force, or **threat** of **force**, or any **persuasion** other than that permitted and defined by law to men who are not unionists, or any other form of **boycotting**, should, in the **opinion** of this Conference, be resisted by **united action**.

This Conference is of the **opinion** that employers should declare that they will not be coerced in the **dismissal** of any labour that has taken **service** with them in the present **emergency**; and in the **event** if any **attempt** being made to coerce such labour to join any trade organization or to interfere with them in the **discharge** of their daily **work**, the combined Associations represented at this Conference will take all possible **means** to insure their personal **safety**.

That this Conference declares that to maintain **discipline**, and thus protect life and property, owners of **shipping** in the coastal and intercolonial trades should not engage or retain in their **employ** any captains or officers who may be members of a Union affiliated with any labour organizations. (Alcott and Alcott, 1980, p. 65)

Texts such as these raise the question of whether nominalization is a question of function or status. It does not seem to be used to organize the report in ways similar to the ways it organizes exposition as outlined above. And one might argue that nominalized language is simply a symbol of literacy and thus education and thus power in our culture. So for resolutions to sound credible, they must be written in language of this kind. This would imply that if we translate written resolutions into spoken English, nothing but status is lost. The problem can be exemplified by taking the second resolution from the text above and asking whether a 'Plain English' translation means the same thing:

This Conference thinks that people should act together to stop unionists attempting to force or threaten to force or illegally persuade non-unionists (to join unions?) or boycotting (businesses employing non-unionists?).

This is a difficult issue which cannot be fully resolved here. It would involve for example looking in detail at *theme* and *information* structure in the English clause, at nominal group complexes, at nominalization and implicitness, at ideology and text structure and so on. But it is important not to lose sight of the instrumental role of abstraction in interpretation. Consider the following passage from Kress (1988b) who is critiquing the recently released NSW *Writing K-12* syllabus:

> It is here where that peculiar theoretical/ideological **mix** of Process Writing in its Australian **manifestation** that I described earlier leads to **inevitable** and **insoluble** problems for a writing curriculum. The **mix** of West Coast psychology ('**ownership**', peer group **work** in the form of '**conferencing**'), romantic liberal/**individualism** (writing as personal '**expression**'), and of a late Leavisite **elitism** (high cultural literary forms as implicit or explicit models of eventual goals, the diary both as **confessional** and as autobiography, the personal narrative as short story or novel, etc) has as its real content the **training** of an individual student subject with a certain kind of **sensibility**. The **view** of **writing** put forward in the document is a view of writing merely as a vehicle for individual **expression**. (Kress, 1988, pp. 14–15)

In the second sentence Kress distills an ideological profile of Process Writing in terms of a mix of West Coast psychology, romantic liberal individualism and late Leavisite elitism. Each of the three ingredients is unpacked to a certain degree in parentheses. But the job of explaining each is really the work of an educational historian. And without abstract language it would be impossible for the historian to range over and make generalizations about the human experience which has been drawn together and interpreted in these terms. Nor could Kress have brought these themes so efficiently to bear on the K-12 syllabus without language of this kind. The main point as far as education is concerned is that students need to learn to read abstract discourse if they are to be functionally literate in our culture and write abstract discourse if they are to interpret their world in a critical way.

The Grammar of Abstraction

What exactly does it mean to make abstract writing 'plain'. Essentially what we are looking at is the relationship between semantics and grammar — between meaning and form. In 'plain' English there is a 'natural' relationship between the two. Actions come out as verbs, descriptions as adjectives, logical relations as conjunctions and so on. These correspondences are outlined below:

SEMANTICS	GRAMMAR
participant	noun
process	verb
quality	adjective
logical relation	conjunction
assessment	modal verb

In the plain English translation of the second resolution given above for example, the processes are realized as verbs. But in the original all these processes except **resisted**, **permitted** and **defined** were coded as nouns:

PROCESS	VERBAL FORM	NOMINAL FORM
'attempt'	attempting	attempt
'force'	to force	force
'threaten'	threaten	threat
'persuade'	persuade	persuasion
'permit'	permitted	permission
'define'	defined	definition
'boycott'	boycotting	boycott
'opine'	thinks	opinion
'resist'	resisted	resistance
'unite'	united	unity
'act'	act	action

The same text includes qualities coded as nouns:

QUALITY	ADJECTIVE	NOUN
'free'	free	freedom
'safe'	safe	safety

and logical relations are expressed in nominal and verbal form: c.f. **in the event of** vs **if**; **insure** vs **so that**.

WRITTEN
... and in the event of any attempt being made to coerce such labour ...

SPOKEN
... if they try to coerce such labour ...

WRITTEN
... the combined Associations represented at this Conference will take all possible means to insure their personal safety.

SPOKEN
... the combined Associations represented at this Conference will do everything they can so that they will be safe.

The last example also contains an assessment of ability, in adjective (**possible**) and modal (**can**) form.

Overall, the effect of abstraction in the grammar of a text is to foreground relational clauses at the expense of material ones and to at the same time foreground nominal groups at the expense of clause complexes. The text itself then codes reality as a set of relationships between things. By way of illustration consider the nominal groups in the Kress passage quoted above:

that peculiar theoretical/ideological mix of process Writing in its Australian manifestation that I described earlier

inevitable and insoluble problems for a writing curriculum

The mix of West-Coast psychology ('ownership', peer-group work in the form of 'conferencing'), romantic liberal/individualism (writing as original personal 'expression'), and of a late Leavisite elitism (high cultural literary forms as implicit or explicit models or eventual goals, the diary both as confessional and as autobiography, the personal narrative as short story or novel, etc.)

its real content

the training of an individual student with a certain kind of sensibility

The view of writing put forward in the document

a view of writing merely as a vehicle for individual expression

As far as relational clauses are concerned, the first nominal group *leads to* (causes) the second, the third *has* (as a part) the fifth, in the role of the fourth, and the sixth *is* (a kind of) the seventh. Kress's critique is formulated as relationships of cause, componence and sub-classification among processes dressed up as things. This brings out the essential continuity between humanities and science as far as interpreting the world is concerned. Both use writing as a tool to analyse the world as if it was simply a collection of thing-like phenomena with various sorts of relationships among them. But whereas the humanities tend to take this process only as far as the interpretations coded in the discourse patterns of the texts, science goes one step further and technicalizes the phenomena and their relationships, translating common-sense understandings into specialized ones. One might say, in summary, that for the historian texts *interpret* the world from a nominal point of view, while for the scientist they *reconstruct* the world as a place where things relate to things.

Chapter 11

Life as a Noun: Arresting the Universe in Science and Humanities*

J.R. Martin

Discourse Technology

Discourses are tools — they do things. That is why they have evolved and thus their functionality determines their character. But because discourses are semiotic tools (and therefore unconscious) they are generally taken for granted in discussions of twentieth-century technology, which focuses instead on designed tools — the material products of conscious invention. Nonetheless, it is the unconscious and evolving discourses of our culture which engender all consciously designed systems. And without a robust interpretation of these discourses, any understanding of the development of material technology in our culture and the ways in which it can be mastered (and masters us) is necessarily incomplete. In this paper consideration will be given to fleshing out the discourses of science and humanities as semiotic technology. The underlying purpose of this deconstruction is to facilitate intervention in the process of literacy development in primary and junior-secondary school (as exemplified in Painter and Martin, 1986; Disadvantaged Schools Programme, 1988; Macken *et al.*, 1989, a, b, c, and d; Rothery 1989a and b).

The context of this discussion then is an educational one. The texts examined will be chosen from what in Australia is referred to as junior-secondary school (Years 7–10, with students typically aged 12 to 16). It is during this period that an attempt is made to apprentice students into the discourses of science and humanities under focus here. Coursework is organized according to discipline, with students moving from one discourse to another throughout the day (Maths, English, History, Geography, Science etc.). Currently in Australian schools no attempt is made to bring the different discourse of these disciplines to consciousness. Indeed, there is a dominant liberal humanist tradition at work which argues that interventions of this kind would be positively harmful (see for example the debates in Reid, 1987 and Threadgold's, 1988 review article of this book).

It needs to be made very clear at this point that the characterization of the discourses of science and humanities developed here will be limited to the

---

* The first six sections of this chapter are taken from *Trends in Linguistics: Functional and Systemic Linguistics*, E. Ventola (Ed), Mouton De Gruyter.

221

Table 11.1:   Four Basic Text Types in Science and Humanities

	'describe'	'explain'
science [technicality]	REPORT [1] (taxonomizing)	EXPLANATION [2]
humanities [abstraction]	REPORT [3] (generalizing)	EXPOSITION [4]

recontextualized pedagogic (Bernstein, 1986) discourses of Australian junior-secondary school. Our work in this area has however shown that these pedagogic texts are very good models of scientific discourses in general (Wignell, Martin and Eggins 1987/1990; Martin, Wignell, Eggins and Rothery, 1988; Shea, 1988; Martin, 1989a on the pedagogic discourse of science in comparison with Halliday, 1987 on the language of physical science). Comparable descriptions of the mature discourses of history are unfortunately not presently available.

Basic Text Types

Because of its focus on text, the discussion will be further restricted to just a few examples, selected as representative of basic features of the discourses of junior-secondary science and humanities. All of the texts considered will be written ones, which is not inappropriate since it is primarily through the resources of written English that disciplines have evolved and differentiated themselves in our culture over the past 400 years (for a discussion of the pedagogic discourse of science in the spoken mode see Lemke, 1985c, 1989, in press, see also Halliday, 1985c, 1989 on the functional differentiation of spoken and written English). The first four texts are from textbooks, which provide the major model of discipline-specific discourses for students in secondary school, who by and large learn to write these discourses by copying directly from these and related reference materials (Wignell, 1987, 1988). Texts 1 and 2 are taken from Shea, 1988.

Provisionally, the four texts can be cross-classified as follows. Texts 1 and 2 are from the field of science, texts 3 and 4 from history. At the same time, texts 1 and 3 have a 'descriptive' function; they give an account of *things as they are*. Texts 2 and 4 on the other hand are 'explanatory'; they attempt to give a reasoned account of *why things are* the way texts like 1 and 3 would present them. The technical terms used to name the genres involved in these tasks are presented in Table 11.1: *report* for the two describing texts, *explanation* for the reason-oriented science and *exposition* for the reason-oriented history. Scientific reports can be distinguished from historical ones with respect to their predominant taxo-nomizing focus, which will be taken up in detail below. The four texts are presented below, divided into ranking clauses (excluding projections which are taken together with the clause projecting them), following Halliday, 1985a. Bold face is used in texts 1 and 2, as it is in the textbooks from which the examples are taken, to highlight technical terms that are being introduced for the first time. To begin, the discussion which follows these texts will focus on **technicality**, which is the predominant discourse feature of the scientific texts, and on **abstraction**, which is foregrounded in the history discourses.

science ('describe')
REPORT: taxonomizing

1. (a) As far as the ability to carry electricity is concerned, (b) we can place most substances into one of two groups. (c) The first group contains materials with many electrons that are free to move. (d) These materials are called **conductors** (e) because they readily carry or conduct electric currents. (f) Conductors are mostly metals (g) but also include graphite. (h) The second group contains materials with very few electrons that are free to move. (i) These materials are called **nonconductors** (j) and are very poor conductors of electricity. (k) Nonconductors can be used to prevent charge from going where it is not wanted. (l) Hence they are also called **insulators**. (m) Some common insulators are glass, rubber, plastic and air. (n) There are a few materials, such as germanium and silicon, called **semiconductors**. (o) Their ability to conduct electricity is intermediate between conductors and insulators. (p) Semiconductors have played an important role in modern electronics. [Heffernan and Learmonth, 1983, p. 212]

Science ('explain')
EXPLANATION

2. (a) If we look at how a tuning fork produces sound (b) we can learn just what sound is. (c) By looking closely at one of the prongs (d) you can see that it is moving to and fro (**vibrating**). (e) As the prong moves outwards (f) it squashes, or compresses, the surrounding air. (g) The particles of air are pushed outwards (h) crowding against and bashing into their neighbours (i) before they bounce back. (j) The neighbouring air particles are then pushed out (k) to hit the next air particles and so on. (l) This region of slightly 'squashed' together air moving out from the prong is called a **compression**. (m) When the prong of the tuning fork moves back again (n) the rebounding air particles move back into the space that is left. (o) This region where the air goes 'thinner' is called a **rarefaction** (p) and also moves outwards. (q) The particles of air move to and fro in the same direction in which the wave moves. (r) Thus **sound** is a compression wave that can be heard. [Heffernan and Learmonth, 1982, p. 127]

history ('describe')
REPORT: generalizing

3. (a) It was mainly after 1965 when the US sent troops into the south on a massive scale, that the movement to have women join the army increased. (b) Women became full-time members of the armed forces in

the south. (c) Many of them held leading positions. (d) About 40 per cent of the regimental commanders of the PLF were women. (e) These were troops who dealt with the American mobile reserves, initiated offensive operations and attacked major US concentrations. (f) All were volunteers who received no salary and when not in combat helped in harvesting, building homes and administering free medical care and medical training.

(g) Women also formed a major part of regional guerilla forces, full-time fighters who operated in the region where they lived. (h) They engaged US forces in the same area by ambushes, (i) encircled bases (j) and attacked posts.

(k) Women in local-self-defence units, or militia women, were not full-time soldiers (1) but fought (m) when their area was attacked, (n) pinning down local forces (o) and keeping their posts permanently encircled.

(p) A higher percentage of women were in the local militia and regional guerilla units than in the PLF. (q) The local militia kept villages fortified with trenches, traps and spikes. (r) These defences were decisive in wearing down the morale of Saigon and US troops. (Simmelhaig and Spenceley, 1984, p. 172)

history ('explain')
EXPOSITION

4. (a) Wars are costly exercises. (b) They cause death and destruction (c) and put resources to non-productive uses (d) but they also promote industrial and economic change. (e) This benefit does not mean that war is a good thing, but that it sometimes brings useful developments.

(f) The Second World War further encouraged the restructuring of the Australian economy towards a manufacturing basis. (g) Between 1937 and 1945 the value of industrial production almost doubled. (h) This increase was faster than otherwise would have occurred. (i) The momentum was maintained in the post-war years (j) and by 1954–5 the value of manufacturing output was three times that of 1944–5. (k) The enlargement of Australia's steel-making capacity, and of chemicals, rubber, metal goods and motor vehicles all owed something to the demands of war. (1) The war had acted as something of a hot-house for technological progress and economic change.

(m) The war had also revealed inadequacies in Australia's scientific and research capabilities. (n) After the war strenuous efforts were made to improve these. (o) The Australian National University was established with an emphasis on research. (p) The government gave its support to the advancement of science in many areas, including agricultural production. (q) Though it is difficult to disentangle the effects of war from other influences, (r) it is clear that future generations not only enjoyed the security and peace won by their forefathers but also the benefits of war-time economic expansion. [Simmelhaig and Spenceley, 1984, p. 121]

Technicality (Science)

As the formatting suggests, scientific discourse is a technical one. This is so because science is concerned with building up an uncommon sense interpretation of the world. To do this it takes common sense as a starting point and 'translates' it into specialized knowledge. The basic semiotic resource available for this translation process is elaboration. At clause rank this meaning is constructed through the relational identifying clause (Halliday, 1985a, pp, 112–28); this is the favoured clause type in scientific discourse for what are commonly referred to as definitions. Sound for example is defined as follows in text 2:

[2r]
sound   is   a compression wave that can be heard
Token   Process   Value

Like all identifying clauses, this clause is reversible (i.e., the technical term can function as either Subject or Complement). Regardless of which way round the clause is the technical term is always Token, and the definition the Value.

[5] (= 2.r reversed)
A compression wave that can be heard   is (called)   **sound**
Value   Process   Token

There are a number of examples of clause rank elaboration is texts 1 and 2; all but 2.r have the Token as Complement, in the unmarked position for New. This makes sense since the technical terms are being introduced in these texts for the first time ('sound' has probably been mapped onto Given in 2.r because of its foreshadowing in 2.b — We can learn just what sound is). The remaining definitions in texts 1 and 2 are as follows:

1.d   These materials are called **conductors**
1.i   These materials are called **nonconductors**
1.l   they are also called **insulators**
1.n   a few materials, such as germanium and silicon, [[*called* **semiconductors**]]
2.l   This region of slightly compressed air moving out from the prong is called a **compression**
2.o   This region where the air goes 'thinner' is called a **rarefaction**
2.r   **sound** is a compression wave that can be heard

Elaboration is also found at group and word rank, once again to translate common sense into specialized knowledge. The structure used here is that of paratactic expansion (which in this context is traditionally referred to as apposition). Formatting is not used to highlight the technical terms elaborated at this rank, which reflects the fact that the elaboration here is not used to define terms but rather to remind readers of the technical way in which scientists talk (the technical term is 'glossed' rather than 'defined' — the structure itself might be elaborated along the lines of 'squashes', or *as we say in science* 'compresses' . . .). Elaboration at group and word rank suggests then that the term can be taken for granted, either because it has already been defined or is not crucial to the discussion at hand. There are three examples of this structure in texts 1 and 2:

1.e   because they readily *carry or conduct* electric currents
2.d   you can see that it is *moving to and fro (vibrating)*
2.f   it *squashes, or compresses,* the surrounding air

Each has the structure 1^= 2, with the technical term following the non-technical one. The structures are outlined below, and related to the analogous clause rank Value ^ Token pattern.

| 1 | carry | squashes | moving to and fro | 'VALUE' |
| =2 | or conduct | or compresses | (vibrating) | 'TOKEN' |

Interestingly, this appositional structure is the only comparable elaborating structure taken up in texts 3 and 4; it is not however used to establish a technical term as will be pointed out below.

3.k   Women in local self-defence units, or militia women,
       1   =2

Abstraction (History)

In sharp contrast to the discourse of science, the discourse of history is not a technical one. Aside from a small set of terms referring to periods of time (the Middle Ages, the Dark Ages, the Renaissance etc.) and possibly some distinctive *-isms* (e.g., colonialism, imperialism, jingoism etc.), relatively few technical terms are used; and where they are used they tend to be borrowed from other disciplines rather than established by historical discourse itself (e.g., socialism, capitalism, market forces, etc.). In this respect the discourse of history closely resembles that of English and it is no accident in Australian secondary schools that teachers of history are commonly English teachers as well, that progressive education has had a deeper influence on the history and English curricula than on any other, and that many of these teachers view structuralist approaches to disciplines as discourses with deep suspicion — commonly denigrating the technicality of these perspectives as ridden with jargon.

The fact that the discourse of history is not technical does not make it any easier to read. The reason for this is that it can be very abstract, especially when explaining why things happened as they did. In linguistic terms this means that reasoning is realized inside rather than between clauses. As an example, consider the following argument from text 4:

4.k   The enlargement of Australia's steel-making capacity, and of chemicals, rubber, metal goods and motor vehicles all owed something to the demands of war.

Within this clause two events are causally related: [event 1: 'Australians fought the war'] *and so* [event 2: 'Australians started making more steel, chemicals, rubber, metal goods and motor vehicles']. But the events are realized nominally rather than verbally — as participants, and the cause-effect relation is realized verbally rather than conjunctively — as a process. In ergative terms, the clause structure is as follows (modelled on more concrete examples such as 'Sue owed $5 to Terry')

4.k	Agent	The **enlargement of Australia's steel-making capacity**, and of chemicals, rubber, metal goods and motor vehicles
	Process	**owed**
	Medium	something
	Beneficiary	to the **demands of war**.

In spoken English this meaning would more likely be constructed as a clause complex, with the events realized verbally and the causal relation expressed as a conjunction. Unpacking the reasoning in 4.k leads to a clause complex such as the following:

[6] (=4.k with reasoning unpacked)
α Australia's steel-making capacity, and of chemicals, rubber, metal goods and motor vehicles enlarged
$^x\beta$ Partly because war demanded it.

Unpacking involves de-nominalizing the participants in 4.k and de-verbalizing the cause–effect relation. The correlation between the 'written' and 'spoken' constructions of this meaning is outlined somewhat more systematically below.

'BURIED REASONING'		'OVERT REASONING'	
Process Medium:	owed something	conjunction:	partly because
Agent:	the enlargement of	Process:	enlarged
Beneficiary:	the demands of	Process:	demanded

Buried reasoning of this kind is a predominant feature of history explanations. There are five further examples in text 4 (the Agent presumed in 4.b–4.e is the activity war):

4.b They cause death and destruction
4.c and put resources to non-productive uses
4.d but they also promote industrial and economic change
4.e . . . it sometimes brings useful developments
4.f The Second World War further encouraged the restructuring of the Australian economy towards a manufacturing basis

In these two introductory sections an attempt has been made to highlight the differences between the discourses of science and history by focusing on technicality and abstraction. At the same time it is important not to lose sight of their continuity. Both technicality and abstraction depend on the same linguistic resource, nominalization — or to put this more generally, on what Halliday, 1985a, pp. 319–45 refers to as grammatical metaphor (see also Halliday, 1967, 1977). Introducing technical terms means placing a Token in relation to its Value, and this entails relating meanings in the grammar as participants. Sound is not a thing, but has to be dressed up as one in scientific discourse in order to be defined. Similarly in history, realizing reasoning inside rather than between clauses means placing an Agent in a causal relation to its Medium, and this entails nominalizing events as participants and verbalizing the logical relation between them. The enlargement of Australia's steel-making capacity and the demands of war are not things, any more than sound is, but they have to be grammaticalized as things in order to reason within the clause. Without grammatical metaphor then, technicality and abstraction would not be possible. And this underlines the significance of writing in the development of discipline — specific discourses — grammatical metaphor is primarily a resource for writing, not speaking. A different kind of consciousness is involved (Halliday, 1985c). Without the technology of writing, science and history as we practice them would not exist.

In the following two sections the function of technicality and abstraction in constructing the fields of science and history will be further examined.

Classifying (in Reports)

One of the distinguishing features of any field, common sense or specialized, is the way in which it classifies experience. Both science and history classify, as can be seen from texts 1 and 3. The history report will be examined first, in order to highlight the non-technical nature of its classification.

Text 3 is a relatively untypical one for history in that it constructs a taxomy (of the roles played by Vietnamese women during the Vietnam War); taxonomy building is the norm for scientific reports, not the exception (Shea, 1988). The key clauses are listed below; 3.a establishes the superordinate to be broken down (bold italics below), and the rest of the report constructs its three sub-classes (bold below).

[SUPERORDINATE]
3. (a) It was mainly after 1965 when the US sent troops into the south on a massive scale, that the movement to have ***women join the army increased*** . . .

[SUB-CLASS 1]
(b) Women became **full-time members of the armed forces** in the south . . .

[SUB-CLASS 2]
(g) Women also formed **a major part of regional guerilla forces**, full-time fighters who operated in the region where they lived . . .

[SUB-CLASS 3]
(k) Women **in local-self-defence units or militia women**, were not full-time soldiers (l) but fought (m) when their area was attacked . . .

Formulated systemically, the taxonomy constructed is as Figure 11.1. Unlike texts 1 and 2, text 3 does not make use of elaborating structures to define technical terms. 'Women **in local self-defence units**' is glossed as militia women,

Figure 11.1: Taxonomy of Army Women (Text 3)

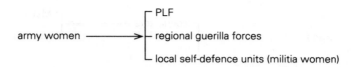

but this is not taken up later in the text as the way to refer to women in this role (i.e., these women are not referred to as **militia women** in 3:p). In 3:p the **local self-defence units** introduced earlier are referred to as the **local militia**, and **regional guerilla units** is used to refer to what were previously described as **regional guerilla forces**. The varied reference to the three different fighting units mentioned in the text is as follows:

the army, PLF, the PLF
regional guerilla forces, regional guerilla units
local self-defence units, militia, local militia, local militia

Thus the terminology used to refer to sub-classes in the taxonomy is not consistent, nor is reference to the sub-classes reduced to single lexical items (the full Classifier^n Thing of the nominal group is retained, as analysed for 3:p below). The history text in other words classifies, but it does not define; the taxonomy it establishes it not a technical one.

> 3:p A higher percentage of women were in the **local militia** and **regional guerilla units** than in the *PLF*.
> local militia   regional guerilla units   P-L-F
>   C   T   C   C   T   C C T

It is important to note here that acronyms such as PLF are not technical terms, but abbreviations. Unlike technical terms, items such as PLF do not have the function of accumulating a number of less specialized meanings in a single lexical item (thus while they may be 'spelled out' through an elaborating structure at group or word rank, they are never defined). Rather, acronyms function as reductions on the expression plane; they make it quicker to write or say a wording — writing or saying P-L-F is faster than pronouncing or spelling the nominal group for which it stands (i.e. . . . .). The proportionalities are as follows:

acronym:expression plane::   (abbreviating sounding/writing)
technical term:content plane   (accumulating meaning)

Acronyms in other words are abbreviations — reductions on the expression plane; technical terms on the other hand accumulate meanings in a single word.

The model of technical terms as accumulations of less specialized meaning is however not adequate. Meaning accumulation is just one aspect of their function. This is because at the same time as they gather together meanings, technical terms also construct new relationships among them. They establish new *valeur*. This

Figure 11.2:   *Taxonomy of Conducting Substances (Text 1)*

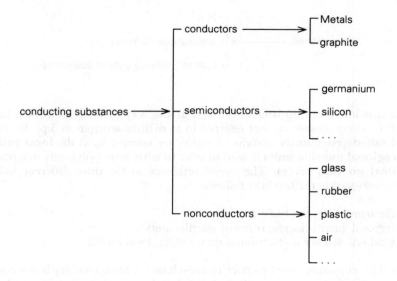

can be seen by looking at the way in which a technical taxonomy is constructed in text 1.

The taxonomy constructed in this text is outlined as Figure 11.2. In this taxonomy, the major sub-classes, conductors, semiconductors and nonconductors are all defined and established in the field as technical terms. At the same time they construct new, uncommon sense relationships among metals, graphite, germanium, silicon, glass, rubber, plastic and air. Technical terms, in other words, both accumulate and change the nature of the meanings they translate into specialized fields. The appropriate metaphor here is not abbreviation but *distillation*. Technical terms, like alcoholic beverages, are both less voluminous products of, and different in kind to, the meanings/materials from which they derive.

It is important to stress that the interpretation of technicality being developed here shows that technical terms are fundamental to specialized discourse. They cannot be dismissed as jargon, because they do not stand in a one-to-one relationship with common-sense terms. The major educational implication of this is that it is utter nonsense to suggest that students learn science better when they are encouraged to use their own words (e.g., Sawyer and Watson, 1987). Science is not science apart from the technical discourse it has developed to reconstruct the world.

As was illustrated above, text 1 relies on relational identifying processes to define its technical terms. But a full range of relational clause resources are brought into play to construct the taxonomy in Figure 11.2. The system network from which these structures are selected is presented as Figure 11.3 below, following Halliday, 1985b, but with intensive Clauses taken one step further in delicacy. The various intensive clause types are illustrated following the network using examples from text 1 (along with one possessive clause type); a

Figure 11.3: Relational Processes in English

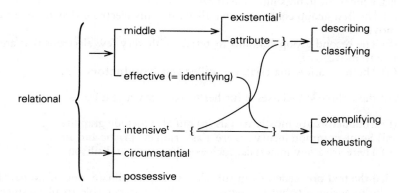

functional interpretation of process and participant relations in these clauses is provided, again following Halliday.

[intensive/existential]
1. n   There   are   a few materials . . .
                     Process   Existent

[intensive/attributive: describing]
1.c   (electrons)   that   are   free [[to move]]
                      Carrier Process   Attribute: adjective [also 1.o]

[intensive/attributive: classifying]
1.f   Conductors   are   mostly metals
       Carrier   Process   Attribute: nominal group [also 1.j]

[possessive/attributive]
1.c   The first group   contains   materials [with many electrons . . .]
       Carrier   Process   Attribute   [also 1.g,h]

[intensive/identifying : exemplifying]
1.m   Some common insulators   are   glass, rubber, plastic . . .
       Value   Process   Token

[intensive/identifying: exhausting]
1.d   These materials   are called   conductors
       Value   Process   Token [also 1.i, 1.n]

The range of relational process types taken up in text 1 reflects the fact that science textbooks build up their taxonomies from superordinate to sub-class rather than the other way round. The text begins with conducting substances, breaking these into two and then three sub-classes:

1. (a) As far as the ability to carry electricity is concerned, (b) we can place most substances into one of two groups.
(c) The first group contains materials with many electrons that are free to move. . . .
(h) The second group contains materials with very few electrons that are free to move. . . .
(n) There are a few materials . . . called **semiconductors**. . . .

Each of these three sub-classes is further broken down as follows:

(f) Conductors are mostly metals (g) but also include graphite. . . .
(m) Some common insulators are glass, rubber, plastic and air. . . .
(n) There are a few materials, such as germanium and silicon . . .

Had the text proceeded from sub-class to superordinate, one relational clause type, the intensive attributive, would have had a larger role to play, since this clause type can be used to assign a sub-class (the Carrier) to a larger class (the Attribute). A text which proceeded in this direction and depended on this particular clause type would read something like the following (the relevant Carriers are in bold face and their Attributes in bold italics in this text):

[7] **Germanium and silicon** are *semiconductors*, because they have an intermediate ability to conduct electricity. **Materials such as glass, rubber, plastic and air** are *insulators*, because they are poor conductors of electricity. **Metals and also graphite**, are *conductors*, because they readily conduct electricity. **Semiconductors, nonconductors and conductors** are *types of substances viewed in terms of their ability to conduct electricity*.

Since it uses relational attributive processes instead of identifying ones, this text does not define technical terms; rather it assumes them. And this reflects the general function of intensive attributive classifying relational processes — namely, to arrange known classes with respect to each other (rather than creating new classes as in defining identifying clauses). This is more appropriate in non-technical than in introductory technical discourse, which explains why this clause type predominates in the history report, but not generally in scientific ones. The examples from text 3 are as follows:

3.b **Women** became *full-time members of the armed forces in the south*
3.d **About 40 per cent of the regimental commanders of the PLF** were *women*
3.e **These** were *troops who dealt with the American mobile reserves* . . .
3.f **All** were *volunteers who received no salary*
3.g **Women** also formed *a major part of the regional guerilla forces* . . .
3.k **Women in local self-defence units, or militia women**, were not *full-time soldiers*

Text 3 also includes descriptive attributive intensive relational processes, which instead of assigning a participant to a more general class, ascribe some quality to it (the first of these, 3.q, is agentive, with the additional function Attributor):

3.q The local militia kept **villages** *fortified with trenches, traps and* . . .
    Attributor   Carrier   Attribute
3.r **These defences** were*decisive in wearing down the morale of* . . .
    Carrier   Attribute

    The group rank elaborations in texts 1 and 3 echo these clause patterns. The elaboration in 1.n is exemplifying, adding a subdivision to the taxonomy being constructed. The elaboration in 3.g on the other hand is classificatory, describing what kind of fighters regional guerilla fighters were.

GROUP RANK (exemplifying)
1.n a few materials,   such as germanium and silicon,
       1   =2
GROUP RANK (classifying)
3.g regional guerilla forces,   full time fighters who operated in the
                            region where they lived
          1   =2

    What these grammatical patterns underline is the fact that whereas scientific reports define, classify and exemplify in order to construct new technical taxonomies, history reports classify and describe in order to generalize across classes of participant, and occasionally, as in text 3, to arrange these generic classes with respect to each other in new ways. The scientific reports in a sense construct new knowledge while the history reports generalize and rearrange the old. Science *invents*; history *interprets* — this at least is how the grammar of their discourse works when the genre focuses on how things are.

Reasoning (When Explaining)

When the way things are is not taken for granted, but explained, very different discourse patterns arise. The change is most dramatic in the science texts, where the Explanation is organized through actions ordered in time. The conjunctive structure of text 2 is outlined in Figure 11.4 to highlight this pattern of organization (conventions as in Martin, 1983). Consequential and temporal relations between events are listed down the right-hand side of the reticulum next to the conjunctions which have (or could have in the case of implicit relations) made them explicit. In text 3, these can be seen to connect almost every pair of clauses. Following Halliday and Hasan, 1976, these relations are referred to as external; they are concerned with organizing events constituting the field. Internal relations on the other hand are concerned with what might be referred to as rhetorical relations within the text itself; they organize the semiosis which materializes simultaneously as the meanings constructing field are made. There are far fewer internal relations in text 2 than external ones.

    Scientific explanations, like reports, give rise to technical terms. Since the technical terms are implicated by a sequence of processes, the distilling function

Figure 11.4: Conjunctive Relations in Text 2 (science)

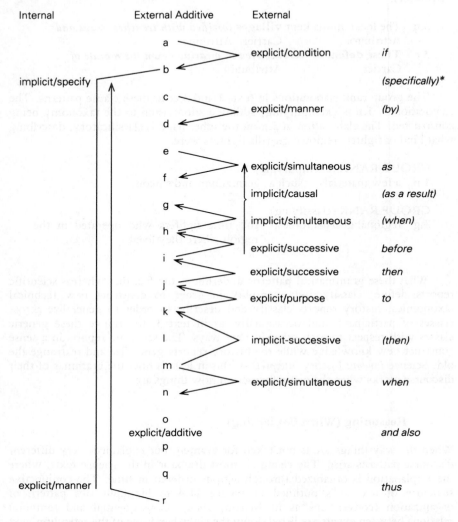

Note: (*conjunctions which could have been used to make implicit relations explicit are enclosed in parentheses.)

of the technicality is striking, with Tokens accumulating a complex set of meanings from their Value (e.g., 2.1 This region of slightly 'squashed' together air moving out from the prong is called a **compression**). This process of distillation culminates in the text with the definition of sound, whose Value makes use of technicality distilled earlier in the explanation (2.r **sound** is a compression wave that can be heard). The process of translating common sense into specialized knowledge is in other words an iterative one, with an indefinite number of steps intervening between a technical term and the every-day 'core' vocabulary through which it might eventually be rendered to an absolute novice.

The rhetorical structure of text 2, as opposed to the temporal and consequential relations organizing its events, is outlined on the left-hand side of the reticulum. In text 2 this structure is a very simple one. Units 2.c through 2.q specify 2.b, explaining step by step how sound is made. Then 2.r 'sums up' 2.c through 2.q — the meaning of 'thus' can be paraphrased 'by specifying how sound is made *in this way . . .*' This internal conjunctive structure means that units 2.c–2.q function simultaneously as the specification of 2.b and the rhetorical means for 2.r. 'Sandwich' structures of this kind are not uncommon in written discourse which seeks to explain.

The conjunctive structure of historical explanations is very different. Here the genre, Exposition, foregrounds internal, not external relations. As is typical of Exposition, very little of this internal conjunctive structure is made explicit; the conjunctions of exemplification, restatement and specification which could however have been used are listed in parentheses in Figure 11.5. Note that the internal relations tend to relate more than two clauses to each other, as in text 2, and that sandwich structures organize 4.e–r and within that structure 4.f-l. The global orientation of internal relations is of considerable significance when interpreting texts with respect to particulate aspects of their generic structure.

Whereas in Figure 11.4, external relations predominate over internal ones, in Figure 11.5 the opposite pattern emerges. This raises the question of whether relations of time, cause and comparison among events are absent in texts like 4, or whether the analysis is simply masking their realization. As has already been suggested in section 3 above, the latter is in fact the case. It is not that causal relations are missing in texts like 4, but that they are realized within, rather than between, clauses, and so not picked up by analyses which focus on conjunctive relations between ranking clauses (unless the text is first stripped of grammatical metaphor, which was not attempted for the analysis represented in Figure 11.5). The pattern of conjunctive relations displayed for text 4 is in fact the result of the abstract nature of historical explanation. And so to interpret the significance of the discourse patterns in Figure 11.5 it is necessary to return to the question of 'spoken' and 'written' grammar in English.

As introduced by Halliday (1985b), grammatical metaphor is the process whereby meanings are multiply-coded at the level of grammar. The argument is that many clauses (and this is especially true in writing) have both a literal reading and one or more others, and that without reading the clause on several levels and understanding the literal reading in relation to the other(s) then the meaning of the clause cannot be fully interpreted. To take an example from text 1, 1.o is literally an intensive attributive relational clause of the descriptive variety (similar to 3.q. and 3.r analysed above). This experiential structure, alongside its textual organization is outlined below.

1.o Their **ability** to conduct electricity is intermediate between conductors and insulators.

'literal'

| Their **ability** to conduct electricity | is | intermediate between conductors and insulators. |

Figure 11.5:   *Conjunctive Relations in Text 4 (history)*

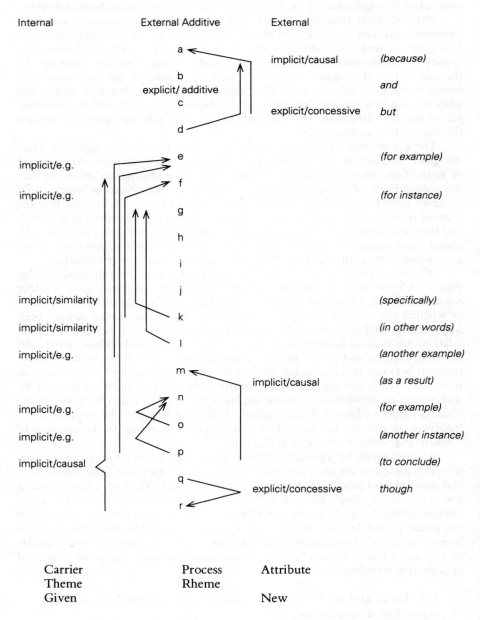

Carrier	Process	Attribute
Theme	Rheme	
Given		New

Beyond this literal interpretation, the clause has at least one further reading in which the nominalization 'their **ability** to conduct electricity' is unpacked. A 'transferred' of 'figurative' reading can then be derived as follows, which notably has consequences for both the clause's experiential and its textual organization:

'transferred'

Semiconductors	**can** conduct	electricity	better than insulators and less well than conductors
Actor	Process	Goal	Circumstance
Theme	Rheme		
Given			New

Although it is not always possible to draw a sharp line between grammatical metaphor and the more traditional notion of a lexical metaphor on which Halliday based his analogy, it is useful to contrast the notion of metaphor in 1.o with that in 4.1. The more traditional metaphor in 4.1 reads literally that war was a green-house. But since war is an activity not a thing and is not used for grow-ing flowers and vegetables in a protected environment it is obvious that the clause must be interpreted on another level as well — one on which war encourages (in opposition to an otherwise debilitating environment) technological progress and change. In order to understand 4.1 then we have to give it more than one reading, beginning with what it actually says. This cannot mean looking through the metaphor to find out what the clause really means, dismissing what it says liter-ally in the process. The fact that we have to read the clause on more than one level is critical — the metaphor makes the clause mean what it does. It is the literal plus (or perhaps better *times*) the transferred reading which counts.

4.1 The war had acted as something of a **hot-house** for technological progress and economic change.

'literal'   war = hot-house
'transferred'   war *x* technological progress and economic change

Proposing that more than one reading is required, raises the question of a base line — of which reading is to be taken as literal. With lexical metaphors, within a given field, this seems unproblematic: a word's 'basic' meaning in that field is taken as literal. Transferred readings are then derived by taking the term's collocational and colligational context into account. In practical terms, dictionar-ies can be used to determine the basic meaning of a lexical item being used metaphorically.

With grammatical metaphor, probably because of the unfamiliarity of this line of interpretation, the question of a base line seems less clear. Halliday's suggestion is that the spoken language of pre-pubescent children be taken as point of departure in reading metaphorical text. And this is a useful suggestion, provid-ing the bias it lends to the interpretation is kept in mind. For one thing the language of these children is closely related to that of mature adults talking in a relaxed and spontaneous way with close friends and family; it is the language speakers return to when intoxicated, stressed or overwhelmed by emotion; it is the language speakers use when they cannot make themselves understood, when talking for example with non-native speakers, young children, or in a noisy room; it is language whose morphology is derivationally simpler than any other, with a lower lexical density and a higher grammatical intricacy (Halliday, 1985a); and so on. The point is that considered along a number of parameters there are ways of motivating the same kind of language as a base line.

Table 11.2:   *Congruent and Incongruent Realizations of Key Semantic Variables*

	congruent	incongruent
ideational: logical conjunctive relation	therefore	reason
ideational: experiential process	advance	advancement
interpersonal assessment	might	possibility
textual reference	he	this (point)

Taking spontaneous spoken discourse as a base line makes it possible to distinguish two types of realization relationship between grammar and semantics which Halliday refers to as congruent and incongruent. A congruent relationship is one in which the relation between semantic and grammatical categories is natural: people, places and things are realized nominally, actions are realized verbally, logical relations of time and consequence are realized conjunctively and so on. In fact, if humans only spoke a language of this kind, there would be no need to distinguish semantics and grammar in the first place; languages could be quite adequately described simply by positing an unstratified content plane.

However, as is well known, the relationship between semantics and grammar is not a simple one. 'Unnatural' relationships are possible; as was outlined above actions can be realized as nouns and logical relations can be realized as verbs. All meanings in fact have more than one manner of realization. A few examples of congruent and incongruent realizations of just four basic semantic categories are outlined, by metafunction, in Table 11.2 (to this point only the ideational variables have been considered). Halliday's (1985a) point about spoken language is that it is the congruent realizations which predominate.

Because of its concern with the fields of science and history, ideational metaphors are the main focus of this chapter. These are further specified in Table 11.3, which outlines congruent and incongruent realizations of four major dimensions of semantic space: conjunctive relations, actions, qualities and participants.

Incongruent realizations are especially frequent in history Expositions as can be seen by highlighting them in bold face in 4m–p below. 'War' will be taken in this analysis as an established technical term not requiring a literal and transferred reading (a dead metaphor in other words); because it is the name of an activity, it could however be unpacked to give a fuller picture of the heavily nominalized nature of texts like 4.

4.m   The war had also revealed **inadequacies in Australia's scientific and research capabilities**.

4.n   After the war **strenuous efforts** were made to improve these.

4.o   The Australian National University was established with an **emphasis on research**.

4.p   The government gave its **support** to the **advancement of science** in many areas, including **agricultural production**.

Table 11.3: Ideational Meaning: Congruent and Incongruent Realizations

Conjunctive relation:			
congruent	cohesive conjunction	therefore	next
	paratactic conjunction	so	then
	hypotactic conjunction	because	before
incongruent	phrasal Process	due to	on
	Process	cause	follow
	Thing	reason	sequel
Action:			
congruent	finite Process	use	deceive
	non-finite Process	using	deceiving
incongruent	Thing	use	deception
	Epithet	useful	deceitful
Quality:			(ATTITUDINAL)
congruent	Epithet	quick	sad
incongruent	Adjunct	quickly	sadly
	Thing	speed	sadness
	Process	quicken	sadden
Participant:			
congruent	Thing	disaster	computer
incongruent	Epithet	disastrous	
	Process		computerize

Table 11.4: Multiple Readings of Ideational Metaphor (in 4.m–p)

	INCONGRUENT [4.m–p]	MORE CONGRUENT [unpacked]	[further unpacking]
QUALITY	inadequacies	inadequate	
	capabilities	capable	[can]
PROCESS	efforts	try to do	
	emphasis	emphasize	[say . . . important]
	research	research	
	support	support	
	advancement	advance	[make better]
	production	produce	
PARTICIPANT	scientific	science	
	agricultural	agriculture	
	strenuous	strain	[to strain]

Incongruent realizations in 4.m–p are aligned with more congruent realizations in Table 11.4. In some cases, the more congruent realizations are themselves incongruent, and can be taken a stage further in the unpacking. This reflects the fact that grammatical metaphor expands the meaning potential of a grammar recursively.

The recursive nature of grammatical metaphor can lead to problems in determining how many readings to give to a highly metaphorical text. One method of reducing the reading process to manageable dimensions is make use of metafunction, unpacking text in three stages. As a first step ideational metaphors can be tackled, unpacking logical metaphors and any experiential metaphors necessary to accomplish this. Returning to example 4.k (The enlargement of Australia's steel-making capacity, and of chemicals, rubber, metal goods and motor vehicles all owed something to the demands of war.) this could proceed as follows:

- i. UNPACKING IDEATIONAL METAPHOR (rendering logical metaphors as conjunctions and unpacking any experiential metaphors necessary to accomplish this)

  α Australia's steel-making capacity **enlarged**, alongside that of chemicals, rubber, metal goods and motor vehicles

  $^x\beta$ **partly because** war **demanded** it.

Subsequently, remaining experiential metaphors could be interpreted, as outlined below:

- ii. UNPACKING REMAINING EXPERIENTIAL METAPHORS

  α Australia's capacity to **make** steel **got bigger**, alongside that of chemicals, rubber, metal goods and motor vehicles

  $^x\beta$ partly because war demanded it

Finally, interpersonal metaphors could be unpacked. In the case of 4.k, it was the people who fought the war, not the war itself, that asked for more steel, chemicals, rubber, metal goods and motor vehicles; and the nominalized modulation 'capacity' can be reworked as a modal verb. The original sentence 4.k has now been rendered as an enhancing clause complex consisting of three interdependent congruent clauses (the metonymy whereby 'Australia' stands for Australians has been left untouched).

- iii. UNPACKING INTERPERSONAL METAPHORS

  α Australia **could** make more steel, chemicals, rubber, metal goods and motor vehicles

  $^x\beta$ partly because **people** demanded them

  $^x\gamma$ **so that they could fight** the war

It can be seen then that history discourse needs to be read on several levels. The buried reasoning introduced in section 4 and highlighted by the conjunction analysis in Figure 11.5 is the result of the process of grammatical metaphor whereby the grammar multiply recodes meanings, breaking down the more natural relations between meanings and their realization which characterizes spontaneous spoken discourse. The result is a highly abstract form of argumentation in which the transitivity resources of the clause take over the reasoning — in metafunctional terms, the experiential subsumes the logical. Writing of this kind is very prestigious, and probably all that is necessary to get a good mark in contexts where exams and essays are marked quickly by staff who are pressed for time. Writing which is not of this kind, but spoken in structure, sounds on the other hand childish and naive. An example of writing of this latter kind will be explored in more detail below.

As far as reasoning in science and history is concerned, it can be seen that compared with history Exposition, reasoning in science Explanations is fairly concrete. Conjunctive relations are realized between rather than within clauses — the text unfolds in a relatively iconic relation to the activity sequence it describes. Where nominalization does occur, it functions to accumulate meanings by subsuming processes as technical terms; it does not function primarily to bury reasoning as in text 4. So whether they are describing how things are or explaining why they are science and history use quite different discourse patterns to construct their fields. Grammatical metaphor is deployed in different ways in the two disciplines, giving very different textures to the apprenticeship students serve in junior-secondary school. Ideationally, technical and abstract discourse are distinctive goings on.

The interpretation to this point has examined the discourse of science and history from the perspective of their field-constructing function. The function of grammatical metaphor in organizing the discourses of these disciplines as text also needs to be considered and will be taken up in the next section.

Packaging (Grammatical Metaphor, Theme and New)

As noted above, grammatical metaphor affects both the ideational and textual structure of the clause. It is not simply a field-oriented resource for burying reasoning or defining terms, but a mode-oriented resource as well — grammatical metaphor is a tool for organizing text. As far as the clause itself is concerned, the critical structures are those associated with the systems of THEME (Theme^ Rheme) and INFORMATION (Given New). These are illustrated for 4.i and 4.n below.

(i)	**The momentum**	was maintained	in the post-war years
	Theme	Rheme	
			New
(n)	*After the war*	strenous efforts were made	to improve these
	Theme	Rheme	
			New

Theme is realized in first position in English and in declaratives conflates with Subject in the unmarked case (only topical Themes will be considered here; marked Themes are realized through Complements and Adjuncts realized before the Subject in declaratives). The analysis of Theme below follows the work of Fries, 1981/1983 and Halliday, 1985a.

Recognizing New is more problematic, since it is realized through tonic prominence and may extend indefinitely leftwards from the tonic syllable (the problem of determining New is further compounded in writing where the tonic is not marked by English graphology). In the analyses which follow clauses will be interpreted as involving unmarked TONICITY, and a minimal domain for the

New will be established by taking New as the highest ranking clause constituent on whose final salient syllable the tonic falls.

Both the systems of THEME and INFORMATION are grammatical resources for relating clauses to their context (including, especially in writing, their co-text). Choices for Theme and New therefore have to be interpreted with respect to the patterns they construct throughout texts. The significance of grammatical metaphor is that it is the grammar's most powerful resource for packaging meanings — for grouping them together into Theme and New. Grammatical metaphor re-textures the clause, allowing it to participate in its context in ways appropriate to the organization of text as text. The interaction of Theme, New and grammatical metaphor can be highlighted through the following analyses of the second and third paragraphs of text 4. Themes are in bold, marked Themes in bold italics and minimal News italicised below:

Theme: unmarked	(**bold**)
Theme: marked	(***bold italics***)
minimal **New**	(*italics*)

(f) **The Second World War** further encouraged *the restructuring of the Australian economy towards a manufacturing basis.*

(g) ***Between 1937 and 1945*** the value of industrial production *almost doubled.*

(h) **This increase** was faster than otherwise *would have occurred.*

(i) **The momentum** was maintained *in the post-war years.*

(j) and ***by 1954–5*** the value of manufacturing output was *three times that of 1944–5.*

(k) **The enlargement of Australia's steel-making capacity and of chemicals, rubber, metal goods and motor vehicles** all owed something *to the demands of war.*

(l) **The war** had acted as *something of a hot-house for technological progress and economic change.*

(m) **The war** had also revealed *inadequacies in Australia's scientific and research capabilities.*

(n) ***After the war*** strenuous efforts were made *to improve these.*

(o) **The Australian National University** was established *with an emphasis on research.*

(p) **The government** gave its support *to the advancement of science in many areas, including agricultural production.*

(q) ***Though it is difficult to disentangle the effects of war*** *from other influences.*

(r) It is clear that future generations not only enjoyed *the security and peace won by their forefathers but also the benefits of war-time economic expansion*

In the second paragraph (4.f–l) grammatical metaphor is strongly associated with Theme. The new information presented in preceding Rhemes is picked up through the nominalizations **increase**, **momentum** and **enlargement**. This interaction of Rheme and Theme is outlined for this paragraph below (nominalizations in bold face):

METAPHORICAL THEMES (second paragraph)

g	Rheme	(the value of industrial productions almost doubled)
h	Theme	This **increase**
h	Rheme	(was faster than would otherwise have occurred)
i	Theme	The **momentum**
j	Rheme	(the value of manufacturing output was three times that of 1944–5)
k	Theme	The **enlargement** of Australia's **steel-making capacity** and of chemicals, rubber, metal goods and motor vehicles

In the third paragraph, grammatical metaphor is more strongly associated with New. As News, these metaphors do not pick up on preceding Rhemes; rather, they have the function of packaging the appropriate combination of meanings as New. A more congruent realization of 4.m for example would manifest *inadequacies in Australia's scientific and research capabilities* as a projected clause, with its own Given New structure. For this clause complex realization the New falls on *revealed* and *inadequate*; but since 4.m introduces Australia's scientific and research capabilities to the text for the first time, this pattern of organization is not really appropriate:

α The war had also *revealed*
β that Australia's scientific and research capabilities were *inadequate*

Paragraph three's actual selections for New are more appropriate and grammatical metaphor is critical to getting the combination of meanings right, as can be seen from the outline below (nominalizations in bold face).

METAPHORICAL NEWS (third paragraph)

(m) **inadequacies** in Australia's **scientific** and **research capabilities**.
(n) to improve **these** [= New of m].
(o) with an **emphasis** on **research**
(p) to the **advancement** of **science** in many areas, including **agricultural production**.
(q) from other **influences**,
(r) the **security** and **peace** won by their forefathers but also the **benefits** of **war-time economic expansion**

Theme and New make different contributions to the organization of text. One way to highlight this is to review the selections for unmarked Theme and New and note the difference in patterning. The choices for text 4 are as follows (nominalizations in bold face):

UNMARKED THEMES	UNMARKED NEWS
War (a,b,d,f,l,m)	costly **exercises**
	death and **destruction**
This **benefit**	to non-**productive uses**
This **increase**	**useful developments**
The **momentum**	the **restructuring** of the . . .

The **enlargement** of Australia's . . .	almost **doubled**
	faster than otherwise would have . . .
The Australian National University	in the post-war years
The government	three times that of 1944–5
	the **demands** of war
	something of a hot-house for . . .
	inadequacies in Australia's . . .
	to improve these [**inadequacies**]
	with an **emphasis** on **research**
	to the **advancement** of science
	. . . from other **influences**
	the **security** and peace won by . . .

The most striking pattern in displays of this kind is the relatively selective nature of thematic choices when compared with New: the list of meanings selected for New in other words is much longer than the list selected as Theme. In text 4, 'war' appears thematically six times, and the lexical set 'benefit-increase-momentum-enlargement' occupies thematic position in four other clauses; the ANU and the government complete the list. Selections for minimal New are more varied and in most cases more complex (with grammatical metaphor deployed to pack several meanings at a time into abstract nominal groups). Theme in a sense provides the text's angle on its field; it is the peg or two on which the rest of the text's meanings are hung. New, by contrast, elaborates the field, developing its meanings — fleshing out the construction of experience with which the text is concerned. Looking upwards to context, Theme is genre-oriented, angling a text in relation to its social purpose; New on the other hand focuses on field, developing the institution at hand. Deployed effectively, grammatical metaphor in writing is a powerful resource for getting a text's angle right and for elaborating its experiential focus in appropriate ways. The textual patterns formed by selections for Theme and New have repercussions for more global aspects of a text's structure. These will be taken up below.

Hyper-Theme and Method of Development

Fries, 1981, 1983, p. 135, in his major paper interpreting the textual function of thematic selection in the English clause, suggests that 'the information contained within the Themes of all the sentences of a paragraph creates the method of development of that paragraph'. Thematic selections in other words articulate an angle on the field which is patterned in semantically definable ways. Themes in texts 1, 2 and 3 can certainly be seen to pattern in specific ways, reflecting the different methods of development appropriate to Reports in science and history and to scientific Explanations.

Fries' examples also indicate a tendency for some texts to predict their method of development. The opposition of wisdom and chance constituting the method of development of his text B for example is predicted by its second sentence: *It* [the English Constitution] *is the child of wisdom and chance*. Similarly the contrast between World War I and World War II is his text H is predicted by that text's first sentence: *Although the United States participated heavily in World*

War I, the nature of that participation was fundamentally different from what it became in World War II. The notion of Theme would thus appear to be relevant at the rank of paragraph as well as clause, with sentences predicting a pattern of clause Themes functioning as point of departure for the paragraph as a whole. The pattern of Themes constituting the method of development of texts 1, 2 and 3 is each predicted by a paragraph Theme in this way.

Fries does not name method of development predicting sentences of this kind, although he does refer to Danes' (1974) patterns of thematic development. In one of these, a single clause Theme predicts a pattern of ensuing clause Themes; Danes refers to Themes of this kind as Hyper-Themes. For Fries' text B and H to have reflected Danes' Hyper-Theme predicting Theme pattern, 'wisdom and chance' in B and 'difference' in H would have to have been thematic in their respective clauses. But the point is that both *wisdom and chance* and *difference* predict their text's method of development equally well whether functioning as Theme or Rheme. Danes' concept of Hyper-Theme in other words seems too restrictive. The notion of paragraph Theme is more appropriate as far as predicting method of development is concerned; and if the notion of Theme is articulated at this rank, the paragraph, then one would expect its realization to be a paragraph constituent (i.e., a sentence or clause) not a clausal one. For these reasons the term Hyper-Theme will be redefined here as a clause (or combination of clauses) predicting a pattern of clause Themes constituting a text's method of development.

In text 1, the Hyper-Theme is the clause complex 1.a–b, which fully predicts the first nine selections for topical Theme (the first 4 of which refer to one group, the conductors, and the next 5 of which refer to the second group, non-conductors). Subsequently the text introduces a third, intermediate group — switching to an existential clause to do so; *there* is very much the appropriate Theme selection at this point since the point of departure of this message is existentiality: 'although it wasn't clearly predicted by the Hyper-Theme (but allowed for by *most substances*), there is in fact another group of substances around.' The text's remaining Themes make reference to these semi-conductors.

HYPER-THEME (text 1)
1.a–b **As far as the ability to carry electricity is concerned, we can place most substances into one of two groups**.

THEMES (rest of text 1)
The first group
These materials
they [these materials]
Conductors
The second group
These materials
Non-conductors
they [non-conductors]
some common insulators
There
Their [semiconductors] ability . . .
Semiconductors

This kind of interaction between Hyper-Theme and method of development is very typical of the Report genre; not surprisingly a closely related pattern is found in text 3. Text 3's Hyper-Theme, 3.a, fully predicts the pattern of Theme selection for the text as a whole: topical Themes refer either to Vietnamese women or the fighting units in which they played their part. It is not in fact clear for some Themes in the first and second paragraphs (i.e., *these, all, they*) whether reference is being made to women, or their fighting units as a whole.

HYPER-THEME (text 3)
3.a **It was mainly after 1965 when the US sent troops into the south on a massive scale, that the movement to have women join the army increased**.

THEMES (rest of text 3)
Women
Many of them [army women]
About 40 per cent of the regimental commanders of the PLAF
These [the PLF/the PLF women?]
All (the PLF/the PLF women?]
Women
They [regional guerilla forces/women in regional guerilla forces?]
Women in local self-defence units, or militia women,
A higher percentage of women
The local militia
These defences [trenches, traps . . .]

Explanations, unlike Reports, focus on an activity; so the Hyper-Theme in text 2 predicts a different kind of Theme selection in which processes have a larger role to play. Processes participate in text 3's method of development in two ways: i. as $^\alpha\beta$ clauses realized before their α in marked Theme position (bold italics below); and ii. embedded as modifiers in nominal groups functioning as Theme (bold below). The text's remaining Themes refer to the air particles that are affected by the movement of the tuning fork's prongs. The text's method of development is thus 'air affected by the motion of the prongs'.

HYPER-THEME (text 2)
2.a–b **If we look at how a tuning fork produces sound we can learn just what sound is**.

THEMES (rest of text 2)
(marked Themes in bold italic, Themes in bold)

By looking closely at one of the prongs,
As the prong moves outwards,
The particles of air
they [the particles of air]
The neighbouring air particles
This region of **slightly 'squashed' together air moving out from the prong**

When the prong of the tuning fork moves back again.
This region **where the air goes 'thinner'**
The particles of air
sound

Texts 1, 2 and 3 can thus be seen to illustrate effectively Fries' point that the pattern of Theme selection in text constitutes its method of development. The various types of method of development reflect in part a text's genre; and there is a tendency for method of development to be predicted by a clause (or combination of clauses) functioning thematically with respect to a number of subsequent clauses. These Theme predictive clauses were referred to as Hyper-Themes. A Hyper-Theme is to a 'paragraph' then, as a Theme is to a clause, keeping in mind that the term paragraph is being used provisionally in this definition to refer to a semantic unit united in part by a predicted and semantically demarcated method of development, which functional unit corresponds with but is not identical to, the graphological unit paragraph in English writing:

Hyper-Theme: 'paragraph'::
Theme: clause

Hyper-New and Point

Fries, 1981/1983, p. 135 distinguishes method of development from two other concepts relevant to the interpretation of paragraphs, topic and point. He characterizes point as the message a text is trying to convey and correlates it positively in his text B with meanings realized through Rhemes and negatively with lexical sets occurring as Theme. Extending Fries' interpretation slightly, it would appear that just as the pattern of Theme selections in a text constitutes its method of development, so the pattern of New selections constitutes its point. And this in turn raises the question of whether the point of a text is in a predictive relationship with a clause (or combination) of clauses that could be defined as Hyper-New. While Hyper-News do seem to be less common in English writing than Hyper-Themes, predictive relations of this kind can certainly be found.

For purposes of illustrating this pattern the domain of the New will be extended to include an additional experiential clause constituent to the left of minimal New. This makes it easier to interpret the sense in which a pattern of News constitutes a text's point and correlates with an ensuing Hyper-New. For text 3, this makes the correlation between the text's point and its Hyper-New, 3.r, quite clear. Text 3's News elaborate the way in which the motion of the tuning fork produces a pattern of movement among adjacent air particles that constitutes an audible wave. These meanings are then accumulated in 3.r as the *compression wave that can be heard* defining sound.

[extended New] minimal NEW (text 3)

[closely] one of the prongs
can see

[is moving] to and fro (vibrating)
[moves] outwards
[squashes or compresses] the surrounding air
[are pushed] outwards
[crowding against and bashing into] their neighbours
[bounce] back
[are . . . pushed] out
[to hit] the next air particles
so on
[is called] a compression
[moves] back
[back] into the space that is left
[is called] a rarefaction
[moves] outwards
[to and fro] in the same direction in which the wave moves

HYPER-NEW
Thus sound is a compression wave that can be heard.

Similarly, in the second paragraph of text 4, the News make the point that industrial production grew dramatically because of the war. The idea of momentum is then picked up metaphorically in 4's Hyper-New, 4.1, with war acting as a *hot-house* for change. Conjunctively, this Hyper-New is in an elaboration relation with 4.g–k (see Figure 11.5 above), which have in turn elaborated the paragraph's Hyper-Theme, 4.f. In spite of this 4's method of development is different from its point: the point of this paragraph is to document how fast things changed—it is the momentum which is constructed as news. In a sense then, the difference in meaning between a paragraph's Hyper-Theme and Hyper-New is a summary of a paragraph's point. This is particularly true where internal relations of elaboration construct a sandwich structure such as that realized through 4.f–l.

[extended New] minimal NEW (text 4 — second paragraph)

[the value of industrial production] almost doubled
[was] faster than otherwise would have occurred
[was maintained] in the post-war years
[was] three times that of 1944–5
[owed something] to the demands of war

HYPER-NEW
The war had acted as something of a hot-house for technological progress and economic change

Hyper-Theme and Hyper-New, as illustrated in above, stand in a complementary relation to each other. Hyper-Themes are prospective—they point forward to a relatively small set of meanings which will pattern as Theme; Hyper-News on the other hand are retrospective—they look backward, ranging across a relatively larger set of meanings patterning as New. Hyper-Themes predict, Hyper-News collect. Taken together they impart a wave-like pattern to a 'paragraph' very similar to that realizing textual meaning in the clause (for the relation of Pike's particle, wave and field theory to metafunctionally differentiated types of structure, see Halliday, 1979); text structure is in other words symbolically related to clause grammar—clause structure is a metaphor for textual organization, and vice versa (see Halliday, 1981, 1982). The analogy developed to this point is outlined below:

Hyper-Theme

Theme	New
Theme	New
Theme	New
. . .	. . .

(Hyper-New)

Macro-Theme and Macro-New

The relationship between Hyper-Theme and Themes and Hyper-New and News raises a question as to the number of levels on which text structure echoes that of the clause. Text 4 provides evidence that the analogy can be taken at least one step further, since in 4 Hyper-Themes are themselves predicted by an introductory paragraph and the point of the second paragraph (which has its own Hyper-New) and the third as well is summed up in 4.q–r. This makes it possible to propose the notion of Macro-Theme, defined as a clause or combination of clauses predicting one or more Hyper-Themes and of a Macro-New, defined as a clause or combination of clauses collecting together one or more Hyper-News. Text 4's manifestation of these patterns is outlined below, with Macro-Theme in bold face, Macro-New in bold italics, Hyper-Themes in small caps and Hyper-New in small cap italics.

Macro-Theme

4. (a) **Wars are costly exercises** (b) **They cause death and destruction** (c) **and put resources to non-productive uses** (d) **but they also promote industrial and economic change** (e) **This benefit does not mean that war is a good thing, but that it sometimes brings useful developments.**

Hyper-theme 1
(f) The second world war further encouraged the restructuring of the Australian economy towards a manufacturing basis.

(g) Between 1937 and 1945 the value of industrial production almost doubled. (h) This increase was faster than otherwise would have occurred. (i) The momentum was maintained in the post-war years (j) and by 1954–5 the value of manufacturing output was three times that of 1944–5. (k) The enlargement of Australia's steel-making capacity, and of chemicals, rubber, metal goods and motor vehicles all owed something to the demands of war.

HYPER-NEW

(1) THE WAR HAD ACTED AS SOMETHING OF A HOT-HOUSE FOR TECHNO-LOGICAL PROGRESS AND ECONOMIC CHANGE.

HYPER-THEME 2

(m) THE WAR HAD ALSO REVEALED INADEQUACIES IN AUSTRALIA'S SCIENTIFIC AND RESEARCH CAPABILITIES.

(n) After the war strenuous efforts were made to improve these. (o) The Australian National University was established with an emphasis on research. (p) The government gave its support to the advancement of science in many areas including agricultural production.

Macro-New

(q) **Though it is difficult to disentangle the effects of war from other influences,** (r) **it is clear that future generations not only enjoyed the security and peace won by their forefathers but also the benefits of war-time economic expansion.**

The interaction of global patterns of *theme* and *information* and *conjunction* is as important at this level of interpretation as it was for the inter-relation of Hyper-Theme and Hyper-New discussed above. Returning once again to Figure 11.5, 4. f–l and 4.m–p are treated as internal elaborations of exemplification in the reticulum which are then taken together as the grounds for the internal relation of clause connected them with 4.q–r. The interaction of exemplification with Macro-Theme gives 4.a–e the meaning of introduction, while the interaction of the causal connection with Macro-New, gives 4.q–r the meaning of conclusion. The systems of internal *conjunction*, global *theme* and global *information* then inter-act to in large part engender the genre's schematic structure. The relationship between the wave-like patterns under consideration here and the particulate structures associated with genre theory will be taken up later.

In longer texts Macro-Themes may themselves be predicted and Macro-News themselves collected. The Table of Contents in a book for example functions as that text's largest Macro-Theme much as the abstract for a scholarly paper functions as its most global Macro-New (albeit somewhat dislocated through its realization in first position in the text; but this can be given a thematic explanation in light of the fact that readers check abstracts to see if the point of the article they are examining makes it worth reading). As Macro-Themes, Tables of Contents tend to be drafted before a book or paper is written, as a kind of plan whereas effective abstracts are written last, after the writers have constructed the new meanings constituting the point of their research. This explains why abstracts submitted to conference organizers are such poor pre-dictors of what so many speakers will actually say and why they are quite un-related to the quality of the talk to be presented (they do however permit paper selection committees to cull papers on ideological grounds, which is the real, though never the professed purpose of vetting papers in the first place).

It can perhaps also be noted here that one common failing in academic writing is to sum up a paper or chapter by writing a 'Macro-Theme' in place of the more appropriate Macro-New. Displaced 'Macro-Themes' of this kind simply reiterate the table of contents, headings, or introductory paragraph of the text without focusing on its news (they could in other words have been as easily understood before the article was read as after). Some texts of course have no point, and so make it difficult to construct an effective Macro-New. This confusion of Macro-Theme and Macro-New has been a problem for educational linguistics in Australia and will be taken up again with respect to text 8 below.

There is in principle no limit to the number of layers of Macro-Theme and Macro-New that can be constructed in text. But Macro-Themes and Macro-News are more common in rehearsed than spontaneous speech, and much more common in writing than talking. Multiple layers of Macro-Theme and Macro-New have to be consciously constructed in writing, through a determined editing and planning process. This is a very important aspect of learning to write, and the kind of problems which arise when it is not taught will be addressed in the next section.

Text Like a Clause

The analogies between text, 'paragraph' and clause outlined to this point can be summarized as follows (where ':' means 'is to' and '::' means 'as'):

Macro-Theme: text::   Macro-New: text::
Hyper-Theme: 'paragraph'::   Hyper-New: 'paragraph'::
Theme: clause   New: clause

Thematic patterns constitute a text's method of development, which correlates in specific ways with genre. Patterns of new information constitute a text's point, which reflects the way in which it constructs its field. Overall, thematic selections are prospective: they make predictions about the thematic structure of text which follows. Selections for new on the other hand are retrospective: they accumulate the new meanings that are made. This interpretation of text as wave is outlined in Figure 11.6.

Year 10 — Writing as you Speak — Text 8

As noted above, these patterns are much more characteristic of writing than speaking, and have to be learned as part of writing across disciplines in secondary school. The writing of students who write as they speak will tend not to reflect an organization of this kind. The following text exemplifies writing without this culminative texture. It was written in a Year 10 geography class in a secondary school in Sydney's inner west. The text is very typical of those composed by students from migrant backgrounds, only a minority of whom will move on from Year 10 to senior-secondary school and university (unmarked topical Themes in bold, marked topical Themes in bold italics and minimal News italicized in 8).

Figure 11.6: *Text as Wave — Local and Global Patterns of Theme and New*

METHOD OF DEVELOPMENT   POINT
(genre focus)   (field focus)

Macro-Theme^n

       Hyper-Theme

predict   Theme . . . New

                                 accumulate

             Hyper-New

               Macro-New^n

WRITING WITHOUT HYPER- and MACRO- WAVES:
8.i (Original 'spoken English' version; 'writing as you speak'; Year 10)

a   **I** think Governments *are necessary*
b   because *if there* wasn't **any**
c   there would be *no law*
d   **people** *would be killing* themselves.
e   **They** help keep our economic system in order *for certain things.*
f   *If there* **wasn't** *no Federal Government.*
g   there wouldn't have been *no one to fix up any problems that would have occurred in the community.*
h   Same *with the State Government.*
i   *if the SG didn't exist*
j   there would have been *noone to look after the school.*
k   **vandalism fighting** would have occurred *everyday.*
l   **The local Government** would be *important to look after the rubbish*
m   because **everyone** would have *diseases*

Writing of this kind is an embarrassment for students, and a real worry for their teachers, who are however hard-pressed to evaluate its shortcomings and teach their students how to do better. Few teachers in Australia have been trained to do more than point to errors in what is commonly referred to as 'grammar, punctuation and usage'. The features of the spoken language that can be identified in this way are easy to 'correct'; this has been carried out for 8.ii below. Editing of this kind makes the text more presentable; but it does not really improve it as a piece of humanities discourse as 8.ii reveals. It changes its status, but not its functionality.

8.ii ('WRITTEN ENGLISH' VERSION; revising 'grammar, punctuation and usage')

a   I think Governments are necessary
b   because if there weren't any

   c   there wouldn't be any law:
   d   people would be killing themselves.
   e   They help our economic system in order for certain things.
   f   If there wasn't any Federal Government
   g   there wouldn't be anyone to fix up any problems that occur in the community.
   h   It's the same with the State Government -
   i   if the State Government didn't exist
   j   there wouldn't be anyone to look after the schools;
   k   vandalism and fighting would occur everyday.
   l   The local Government is important to look after rubbish,
   m   because otherwise everyone would have diseases.

The reason for this is that editing for 'grammar, punctuation and usage' does not improve the text's patterns of Theme and New; effective writing is not just a question of good manners — etiquette is not enough. Topical Themes were underlined in 8.i above (unmarked Themes in bold face and marked Themes in bold italics) and News italicized. These selections are listed below for each ranking clause in the text (version 8.ii).

UNMARKED THEME	UNMARKED NEW
I	necessary
-*	weren't
-	any law
people	would be killing
They [governments]	certain things
-*	any Federal Government
-	the community
It [i.e the argument]	the State Government
-*	didn't exist
-	the schools
vandalism and fighting	everyday
The local Government	rubbish
everyone	diseases

MARKED THEME
if there weren't any
if there wasn't any Federal Government
if the State Government didn't exist

[* Following Fries, 1983, Theme is not analysed in clauses which themselves function as marked Themes nor in the α clauses they function as marked Theme for; these clauses have however been analysed for News]

As with 'grammar' punctuation and usage', text 8's method of development is a very spoken one. The argument is organized around three marked Themes realized by enhancing β clauses, each of which is an existential clause with negative polarity; the writer suggests that Governments are necessary because of what would and wouldn't happen if there weren't any and uses the grammar of spoken English to organize his text along these lines. The text's method of development is not reflected in its choice of unmarked topical Themes, nor is it predicted in any way by a Hyper-Theme; neither is the text's point accumulated in a Hyper-New.

One way of working towards a more 'written' version of text 8 is to revise *theme* and *conjunction*, making levels of government the text's method of development and using internal conjunction to connect its arguments. These changes can be made without affecting the text's experiential meaning, as 8.iii illustrates below. Selections for unmarked Theme and New are listed following the text (bold face highlights the internal conjunctive structure which has been added to the text).

8.iii (RE-ORGANIZED VERSION; revising theme and conjunction)

a   I think Governments are necessary at different levels for a **number of reasons**.
b   They make laws, without which people would be killing themselves,
c   and help keep our economic system in order.
d   **To begin**, the Federal Government fixes up problems that occur in the community.
e   **Similarly**, the State Government looks after schools,
f   preventing vandalism and fighting.
g   **Finally** the Local Government is important to look after rubbish:
h   otherwise everyone would have diseases.
i   **As a result of these factors**, Governments at several administrative levels are necessary.

UNMARKED THEME	UNMARKED NEW
I	a number of reasons
They [governments]	would be killing
the Federal Government	in order
the State Government	the community
the Local Government	schools
Governments	vandalism and fighting
as a result of these factors	important to look after rubbish
	diseases
	necessary

Further improvements can be made by working on Hyper-Theme and Hyper-New, which involves introducing a number of grammatical metaphors (underlined in 8.iv below). In 8.iv the movement of the text through governments at various administrative levels is predicted by 8.iv.a–b (its Hyper-Theme) and the text's point is accumulated in 8.iv.h (its Hyper-New).

[8.iv] (adding Hyper-Theme and Hyper-New)

a   I think Governments are **necessary** for a **number** of **reasons**.
b   These have to do with the special *duties* of Governments at different **administrative, levels**-Federal, State and Local.
c   To begin the Federal Government fixes up *problems that occur in the community* . . .

d   Similarly the State Government looks after *schools*;
e   this prevents *vandalism and fighting* . . .
f   Finally the Local Government is *important to look after rubbish*:
g   otherwise everyone would have *diseases* . . .
h   As **a result** of their **concern** with general **problems, education** and waste **disposal**, Governments at several **administrative levels** are **necessary**

The next step would be to develop the arguments in 8.iv.c, 8.iv.d–e and 8.iv.f-g as paragraphs, which would have the effect of transforming 8.iv.c, 8.iv. d–e and 8.iv.f-g into Hyper-Themes and 8.iv.a–b into a Macro-Theme. This development will not be pursued here. The point of the exercise has simply been to demonstrate that improving writing depends on changing texture, not spelling, punctuation, and usage and that writing can be improved without rejecting an apprentice writer's 'ideas'. The experiential content of 8.iv is very like that of 8.i; it is the textual structure which has changed.

It needs to be stressed at this point that there are those who would cherish version 8.i and others like it, celebrating its difference from the writing towards which it has been moved 8.iv. Their response would apparently involve sitting down with the child to value 8.i's meanings, working together with the student on what he really wanted to say, helping him express himself—in his own words—so that he could feel some sense of ownership of the text. The response would be passively benevolent, not interventionist; the student's meanings would be respected at all costs. And care would be taken not to actually teach the student about writing, since real understanding can only come from what writers figure out for themselves, not from what they might be told.

It is difficult to imagine anything more childish and fundamentally patronizing than an attitude of this kind (a manifestation of the liberal humanist rhetoric used by middle-class educators to oppress generations of migrant, aboriginal and working-class students in Australia). And it is safe to say that in many secondary schools in Sydney's inner west, the approach would be recognized for what it is. There are students who would be quite open about telling anyone putting them down in this way to 'get stuffed'. The text, along with most of the others written in the class, is drivel—the students know it and are embarrassed by it; their teachers know it and are desperate for something which will help. These writers are leaving school at the end of the year to look for work in a community that expects them to have learned to write; and they are entering a workforce which depends increasingly on exchanging information through sophisticated computer based technology. Love and understanding are just not relevant. As at least one Disadvantaged Schools Programme literacy consultant has found when arriving at schools of this kind, the first response encountered is as follows: 'If you're coming here with any more process writing, forget it; *it doesn't work with these kids.* Don't waste our time!'

Fortunately, there is no need to waste time. Writing can be taught. And a few hours work on genre, theme and nominalization in the context of humanities writing of this kind can improve the situation tremendously. During 1988 Joan Rothery worked with Howard Luckman, a History teacher in a secondary school adjacent to that attended by the writer of 8.i which attracts large numbers of students from a similar migrant background. Text 9 is an example of an average

Exposition written in a Year 10 class. As one of the students commented to Joan Rothery about the intervention: 'Gee Miss, it's good to have someone tell you how to make your work better'. After the years of progressive education, it came as something of a surprise to students that improvements in writing could be taught, rather than urged, however benevolently.

Year 10 — History Exposition — Text 9

There are many reasons to support the idea that History is a worthwhile subject to study. The following are: the importance of learning about the past, the relevance of history to a career and, last but not least, that history is interesting.

THE FIRST REASON IS THE IMPORTANCE OF LEARNING ABOUT THE PAST. First of all, by understanding history we can understand the present. For instance, how the human race has proceeded the way it has, from early humans through the ages (e.g., the Ice-Age, Stone Age). These factors give evidence of why we have changed and what we might look like in the future. In addition, by comprehending the past we might learn enough not to repeat the same mistakes. For example, if one's ancestors built a house near a river and the river over-flowed and killed some of them, we would hope a lesson would be learned about not building in that place again.

ANOTHER FACTOR IS THAT HISTORY CAN BE RELEVANT TO A CAREER. History can thus be included in a profession such as archaeology. Further, average people trying to trace their family trees as far as they possibly can can find history useful. Also, History is important if you are good at it. Those having a natural talent in history should feel confident enough to develop it.

FINALLY, HISTORY IS INNATELY INTERESTING. New things can be interesting for students. For example, a student could sit and think about how the pyramids were built, grueling work day by day, as their construction and how they retained their shapes . . . History covers many types of things, such as different buildings, cultures and customs . . . Another factor is that there are two types of history. One is written history, with documents, such as diaries, which can be discovered and studied . . . The other is oral History. For example, a Prime Minister can make a remark which will affect the activity of his country and . . . maybe the world.

Therefore, the following reasons support the contention that the study of History is worthwhile: the importance of learning about the past; the relevance of History to careers; and, lastly, the innate interest of History. These reasons support the idea that History is a worthwhile Subject to study.

Text 9 has a clear pattern of Macro-Theme (the first paragraph) and Hyper-Themes (formatted in small caps in text 9), and makes use of a number of nominalizations to articulate these (e.g., *the importance of learning about the past, the relevance of history to a career*). Where the text falls down is in its failure to distinguished Macro-Theme from Macro-New (the last paragraph). The last paragraph simple reiterates the first (note however the nominalization of the Macro-Theme's projection *that history is interesting* as the Macro-New's *the innate interest of History*). The reason for this is that the student learned too well; the educational linguists involved in this project (Martin and Rothery) had failed to recognize the difference between Macro-Theme and Macro-New at this stage of their research. Macro-News need to review a text's point, collecting meanings that could not be previewed in a Macro-Theme because they had not yet been made. Practice rebounded on theory; next time round, the teaching will improve.

Genre: Wave as Particle

As noted above for text 4, it was the interaction of conjunction relations with Macro-Theme and Macro-New which played a key role in constructing an introduction-body-conclusion progression for that exposition. Across genres, there is a similar correlation between global patterns of theme and information and the goal oriented schematic structures derived from genre in the contextual model assumed here (see Martin, 1984a, 1985a for a brief introduction). These complementary perspectives on text as particle (genre) and text as wave (mode) are summarized in Table 11.5 for the genres represented in texts 1 to 4. The schematic structures for scientific reports and explanations are taken from Shea, 1988. For exposition, see Martin and Peters, 1985; Martin, 1985b, 1989; Kalantzis and Wignell, 1988; for the structure of non-technical reports see Eggins *et al.*, 1987; Rothery, 1989a; Macken *et al.*, 1989.

Shea's schematic structure for scientific reports begins with a Cue which signals that taxonomy construction is underway and allows for embedding of further reports and explanations in the Taxonomic segment which follows the Cue (embedded reports may develop further delicacy in the taxonomy). In text 1, 1.a–b realizes the Cue, 1.c–p the Taxonomic segment. Seen from the perspective of textual meaning, the Cue is a Hyper-Theme.

The schematic structure for history reports is not taxonomy focused. These reports begin with a General Statement followed by an indefinite number of Descriptions. Because reports are so strongly oriented to field, it is difficult to make more specific predictions about their staging. Text 3 was oriented to classifying the combat roles played by Vietnamese women in the Vietnam war; but this is just one of the many semantic strategies used by history reports for organizing information. Clause 3.a realized the General Statement in text 3 and also functioned as its Hyper-Theme. This was followed by three descriptive elements, 3.b–f, 3.g–j, 3.k–r.

According to Shea, scientific explanations begin with a very general Link stage, which in science textbooks bridges between the explanation and the report into which it is embedded. Following this there is an indefinite number of Implication Sequences, any one of which may optionally be explicitly closed with a stage called State. In textual terms, the Link functions as Hyper-Theme and the

Table 11.5:   Structures of Texts

GENRE schematic structure (particle)		MODE culminative structure (wave)
REPORT (classifying — science; from Shea, 1988)		
Cue		Hyper-Theme
Taxonomic segment ([[report]]) ([[explanation]])		
REPORT (generalizing — history)		
General statement		Hyper-Theme
Description^n		
EXPLANATION (science; from Shea, 1988)		
Link		Hyper-Theme
Implication Sequence^n	Processes	
	(State)	Hyper-New
EXPOSITION (humanities)		
Thesis		Macro-Theme
Argument^n	Assertion	Hyper-Theme
	Elaboration	
	(Re-assertion)	[Hyper-New]
Reiteration		[Macro-New]

State as Hyper-New. Text 2 has 2.a–b as Link, the Implication Sequence 2.c–q and reaches closure with the State 2.r. As 2 exemplifies, arresting the implication sequence to define technical terms associated with processes is a common function of the State element of structure.

Exposition typically has a somewhat more elaborate particle and wave structure. The genre's Macro-Theme introduces a Thesis (4.a–e in text 4) for which an in principle indefinite number of Arguments will be advanced. Each Argument then has its own potential Assertion^ Elaboration^ Re-assertion structure, with Assertion as Hyper-Theme (e.g., 4.f in 4) and Re-assertion as Hyper-New (e.g., 4.l in 4). Finally, the position advanced in the Thesis is rearticulated—a Reiteration stage which functions simultaneously as the text's Macro-New (4.q–r in 4) As noted above, the label Reiteration (and the label Re-assertion as well) can be misleading unless the difference between method of development and point is taken into account. The interpretations of text as particle and wave are complementary—neither perspective is sufficient, as literacy interventions based simply on particulate models of generic structure have shown.

Anti-Discourses (Subversive or Pathological?)

The function of grammatical metaphor in constituting the pedagogic discourses of science and humanities has now been considered from the perspectives of field (technicality and abstraction), mode (Macro-Theme/New, Hyper-Theme/New) and genre (the correlation of particle with wave). In closing, the texture constitutive of these two semiotic resources will be foregrounded by considering two discourses which actively resist the drift of their semiosis. The first discourse to be considered will be ethnomethodology—an anti-technical discourse. The second discourse deconstructed will be post-structuralism, as practiced by Derrida; this discourse is very abstract, but certainly frustrates the expectancies of readers accustomed to 'rational' thrust of the abstraction in the humanities texts discussed above.

The discussion of ethnomethodology is based on Eggins (1986) who studied the refusal of ethnomethodologists to elaborate a technical discourse characteristic of science and social science. Ethnomethodology's resistance to technicality arises from their wish to distance themselves from the discourses of mainstream American sociology (whose theories and quantitative methods they regard with great suspicion; see Levinson, 1983, pp. 294–5). The writing of ethnomethodologists symbolizes their concern not to impose categories on social interaction but to uncover the techniques which speakers themselves use to construct their social relations through everyday conversation. The resulting discourse, though non-technical, is extremely abstract. In the following text (from Eggins' study), action verbs are almost completely eliminated. Instead of writing for example that 'speakers minimize gap and overlap', ethnomethodology prefers that '**minimization** of gap and overlap be *accomplished*'. And even lexically empty verbs like *accomplish* may themselves be nominalized (e.g., the **accomplishment** of turn-transfer). This preference for realizing action as a noun and using empty verbs as processes can be illustrated from text 10 as follows:

> minimization of gap and overlap is *accomplished*
> the use of the allocational technique has been constructionally *accomplished*
> 'current speaker selects next' techniques may be *accomplished*
> the accomplishment of turn-transfer does not *occur*
> those ['current selects next' techniques] may be *applied*
> self-selection may not be *exercised*
> self-selection is not *done*
> self-selection . . . must be *done*

The next also nominalizes a number of these 'emptied verbs':

> The **use** of the allocational technique
> the **accomplishment** of turn-transfer
> the **use** of self-selection techniques
> the **non-use** of 'current selects next' techniques

The resulting text is one that takes several readings to unpack. Grammatical metaphors are in bold face in text 10 below to highlight its abstraction. In addition, the text's empty verbs are in italics.

ETHNOMETHODOLOGY (from Eggins, 1986)
'anti-technicality'

> [10] **Minimization** of *gap* and *overlap* is *accomplished* in two **ways**: one
> **localizes** the **problem**, the other **addresses** it in **localised** forms. The
> *rule-set*, along with the **constraints** imposed mutually by the **options**
> in it, eliminates *gap* and *overlap* from most of **conversation** by
> eliminating *gap* and *overlap* from most single *turns*. The *rules* provide
> for *turn-transfers* *occurring* at *transition-relevance* places, wherever the **use**
> of the *allocational technique* has been **constructionally accomplished**.
> Thus 'current speaker selects next' **techniques** may be *accomplished* at the
> very **beginning** of the *unit-type* employed in a *turn* (e.g., by the **use** of
> an *address term* for certain *unit-types*); but the **accomplishment** of
> *turn-transfer* does not *occur* until the first **possible** *transition relevance*
> place. The **use** of *self-selection techniques* is **contingent** on the **non-use**
> of 'current selects next' **techniques**, and those may be *applied* at any
> point up to the first *transition-relevance* place, hence *self-selection* may
> not be *exercised* (the **technique** selected or the **transfer** attempted) until
> the first *transition-relevance* place. *Current speaker* may continue (*rule*
> *1c)* if *self-selection* is not *done*, thus recycling the *rules*: hence *self-*
> *selection* to be *assured*, must be *done* AT the *transition-relevance* place.
> The *turn-taking rule-set* thus provides for the **localization** of *gap* and
> *overlap* **possibilities** at *transition-relevance* places and their immediate
> **environment**, so that the **rest** of a *turn*'s 'space' is **cleansed** of **system-**
> **atic bases** for their **possibility**. (Sachs, Schegloff and Jefferson 1974, pp.
> 705–6)

It might be expected from the characterization of writing like text 10
given to this point that ethnomethodology has replaced the discourse of science
with that of the humanities. However this is not the case. Ethnomethodogists
are certainly structuralists, and their discourse focuses on building elaborate
classifications of their data. The special language used to classify their data has
been highlighted in bold face in text 10 but is different from the technical lexis
considered above. A few simple specialized terms are found, which look
disarmingly like everyday words (e.g., '*gap, overlap, turn* and *rule*'). But in general
ethnomethodologists seem unhappy about distillation. Rather than accumulating
meaning in a single term through definition, they prefer to retain the 'semantic
history' of a category through compound words (e.g., '*rule-set*') and Classifier^
Thing structures (e.g., '*allocational technique*') , some of which have compound
(e.g., '*self-selection techniques*') or embedded (e.g., '*current speaker selects next techni-*
ques') Classifiers. A complete breakdown of this kind of categorization in text
10 is provided below:

specialized THINGS:

(simple)
gap, overlap, turn, rule

(compound)
rule-set

turn-transfers
transition-relevance
unit-type
self-selection
turn-taking

Classifier^ Thing:

(simple Classifiers)
allocational technique
address term
current speaker

(compound Classifiers)
turn-taking rule-set
self-selection techniques
transition-relevance place

(embedded clause classifiers)
'current speaker selects next' techniques
'current selects next' techniques

One result of this kind of refusal to distill is a discourse which is so grammatically metaphorical that it is next to impossible to read. Another is that readers, and apparently the ethnomethodologists themselves (see Eggins, 1986, on the problem of gaps on their taxonomies), have trouble reconstructing from their discourse the classification they are elaborating. This pathology clarifies an important aspect of the function of technical lexis in scientific discourse, which is to rid the discourse of the grammatical metaphors which were essential to the process of constructing a scientific reading of reality in the first place.

In text 2 for example, the technical term **compression** was defined as follows:

[cf. 2.1 *This region of slightly compressed air moving out from the prong is called a* **compression**]

This meant that in 2.r Heffernan and Learmonth could define sound as '*a compression wave that can be heard*'. An ethnomethodologist would have been more likely to refer to an '*air-compression wave*' than a '*compression wave*'; and would probably have resisted defining a term like **sound** in the first place, preferring to refer to '*audible air-compression waves*' or more likely the '*accomplishment of audible air-compression waves*' (or even '*the occurrence of the accomplishment of audible air-compression waves may have the effect of . . .*' perhaps leading to '*audible air-wave accomplishment occurrence effect*' in place of *sound*). Scientific discourse on the other hand proceeds by treating *sound* as a non-metaphorical participant, greatly reducing the burden on readers as far as unpacking layers of grammatical metaphor are concerned. Providing the reader has digested what sound is, this is a great advantage as the discourse proceeds to distill further technical terms.

In some respects one might expect that in refusing the resources of technicality, and developing a confusing and ponderous anti-discourse such as that illustrated in text 10, that ethnomethodology was doing itself a disservice. Their discourse frustrates the development of categories which would make contact with mainstream sociology, grounding macro-theory in micro-interaction; and this has certainly kept ethnomethodologists on the margins of their discipline. Beyond this, their refusal to deal with with macro-categories such as the discourses of class, gender, race and age which have been the focus of considerable attention in twentieth century critical theory makes much of their analysis problematically naive (see for example Thompson's, 1984, pp. 117–18 critique of Sach's damagingly gendered analysis of a dirty joke).

Outside sociology however, in the field of linguistics, their work on conversational structure has very high status. Levinson (1983) for example, in his textbook on pragmatics, briefly dismisses systemic work on conversation (7 pages) and then develops a long and adulatory introduction (76 pages) to the work of Sachs, Schegloff, Jefferson and their co-workers. There are probably two factors at work in reactions of this kind. One is that the ethnomethodologist's discourse appeals to naive empiricism—Sachs, Schegloff and company apparently stay close to their data and never posit more than can be shown to be there. The other factor is that the discourse of ethnomethodology makes discourse look very different from syntax–analysis is at a preliminary stage, the phenomena considered are hard to deal with, common-sense sorting is the best that can be done at this stage, and so on. All of this sits very nicely with the hegemony of autonomous syntax in American linguistics and with a linguistics that is not designed to be socially accountable in any direct way: 'we're not really ready for discourse yet (nice that some people are working on it though); better to stick with the tried and true.' In short, what is disfunctional at one level (field) may be very functional at another level (ideology); the success of the anti-discourse of the ethnomethodologists has to be interpreted in political terms.

The second anti-discourse to be considered here is that of post-structuralism. Unlike ethnomethodology, post-structuralism makes no attempt to develop an elaborate categorization of the world. As practised by Derrida, its data in fact is other people's theories — which it attempts to deconstruct. The discourse it uses to do so is not a technical one; only a few specialized terms are developed (e.g., '*transcendental signified, differance, grammatology* etc.) and these do not combine to form elaborate new systems of classification (contrast the taxonomically oriented work of Habermas, 1984). Rather, the main thrust of the semiosis lies in its argumentation. Post-structuralism is a humanities discourse, not a scientific one.

Derrida himself is of special interest to systemicists because of his deconstructions of Saussure (as Hasan, 1987b, points out the structuralism that post-structuralism is post- to is that articulated by Saussure; most of the rest of twentieth-century linguistics is completely ignored). In particular Derrida is interested in assumptions about spoken and written language which permeate linguistic theory and he apparently takes Saussure to task for laying undue emphasis on the primacy of the spoken language. It is not clear whether Derrida fully appreciates the extent to which linguistic theory is predicated on and shaped by the existence of writing (Derrida, 1974, p. 27); and at least one linguist has argued that Derrida has thoroughly confused the relation between spoken and written language (Hall, 1987). Whatever the case, it is the anti-rationality of

Derrida's discourse, not whether he is right or wrong about Saussure and writing, which is at issue here.

Extracts from Derrida's critique of Saussure are presented as text 11 below. Reading anti-discourses is heavy going, and like the ethnomethodologists, Derrida can be very confusing. In 11 Derrida's rhetorical strategy is to quote from Saussure and then to paraphrase aspects of these quotations in his own words. The elaboration of Saussure's position is strongly attudinal; Derrida repeatedly takes Saussure's comments, often out of context, and systematically exaggerates their original evaluative force. The effect is to mock Saussure, ridiculing his position as absurd. One of the main problems reading texture of this kind has to do with projection. It is not always clear when Derrida is speaking, when he is paraphrasing and when he is quoting Saussure (quotation marks are used but quotations, however long, are not set apart from the rest of the text). In 11, quotations from Saussure have been formatted in lower caps, and Derrida's evaluative paraphrases are in bold face in order to help keep the voicing clear.

POST-STRUCTURALISM
anti-'rationality'

[11] "WRITING, THOUGH UNRELATED TO ITS INNER SYSTEM, IS USED CONTINU- ALLY TO REPRESENT LANGUAGE. WE CANNOT SIMPLY DISREGARD IT. WE MUST BE ACQUAINTED WITH IS USEFULNESS, SHORTCOMINGS, AND DANGERS." Writing would thus have the exteriority that one attributes to utensils; to what is even an **imperfect** tool and a **dangerous, almost maleficent**, technique . . . It is less a question of outlining than of protecting, and even of restoring the internal system of language against the **grayest, most perfidious, most permanent contamination** which has not ceased to **menace**, even to **corrupt** that system . . . The **con- tamination** by writing, the fact of the **threat** of it, are **denounced** in the accents of the **moralist** or **preacher** by the linguist from Geneva. The tone counts; it is as if, at the moment when the modern science of logos would come into its autonomy and scientificity, it became neces- sary again to attack a **heresy** . . . Thus **incensed**, Saussure's **vehement** argumentation aims at more than a theoretical error, more than a **moral** fault: at a sort of **stain** and primarily a **sin** . . . For Saussure it is even a garment of **perversion** and **debauchery**, a dress of **corruption** and disguise, a festival mask that must be **exorcised**, that is to say **warded off**, by the good word: "WRITING VEILS THE APPEARANCE OF LANGUAGE; IT IS NOT A GUISE FOR LANGUAGE BUT A DISGUISE." . . . "WHOEVER SAYS THAT A CERTAIN LETTER MUST BE PRONOUNCED IN A CERTAIN WAY IS MISTAKING THE WRITTEN *IMAGE* OF A SOUND FOR THE SOUND ITSELF . . ." . . . A **dangerous promiscuity** and a **nefarious complicity** between the reflection and the reflected which lets itself be **seduced narcissistically** . . . "THE LITERARY LANGUAGE ADDS TO THE UNDESERVED IMPORTANCE OF WRITING. . ." When linguists become **embroiled** in a theoretical mistake in this subject, when they are **taken in**, they are *culpable*, their **fault** is above all *moral*; they have yielded to imagination, to sensibility, to **passion**, they have fallen into a 'TRAP . . . THE LANGUAGE DOES HAVE A DEFINITE AND STABLE ORAL TRADITION THAT IS INDEPENDENT OF WRITING, BUT THE INFLUENCE

[PRESTIGE] OF THE WRITTEN PREVENTS OUR SEEING THIS.' We are thus not blind to the visible but **blinded** by the visible, **dazzled** by writing . . . For Saussure, to give in to the 'PRESTIGE OF THE WRITTEN FORM' is, as I have just said, to give in to *passion*. It is **passion** — and I weigh my word — that Saussure analyses and criticises here, as a **moralist** and psychologist of a very old tradition. As one knows, **passion** is **tyrannical** and **enslaving**: 'PHILOLOGICAL CRITICISM IS STILL DEFICIENT ON ONE POINT: IT FOLLOWS THE WRITTEN LANGUAGE SLAVISHLY AND NEGLECTS THE LIVING LANGUAGE . . . THE TYRANNY OF WRITING. . .' . . . That **tyranny** is at bottom the mastery of the body over the soul, and **passion** is a **passivity** and **sickness** of the soul, the **moral perversion** is *pathological* . . . **perverse cult . . . sin of idolatry** . . . The **perversion of artifice** engenders **monsters** . . . Writing . . . participates in the **monstrosity** . . . It is a **deviation from nature** . . . Saussure's **irritation** with such possibilities **drives** him to **pedestrian** comparisons . . . (Derrida, 1974, pp. 30–9)

Once the projection is sorted out, there can be no doubt that Derrida's representation of Saussure's attitude to speaking and writing is seriously distorted. Few who had read the *Cours*, examined the quotes in context, and kept in mind that the record of Saussure's ideas we are working with was assembled from students' notes could consider Derrida's representation of Saussure's position a scholarly one. And there are many who would be offended by Derrida's moralistic mocking of Saussure in this way. Setting these objections aside however (allowing perhaps for cultural difference), there is a more troubling aspect of Derrida's writing having to do with Macro-New. Having worked through passages like that represented in text 11, one expects a summary statement which brings the ridicule to a head. Instead, Derrida presents what appears to be a complete contradiction; the actual Macro-New Derrida offers for text 11 is as follows:

MACRO-NEW
[12] I hope my intention is clear. I think Saussure's reasons are good. I do not question, on the level on which he says it, the truth of what Saussure says in such a tone. And as long as an explicit problematics, a critique of the relationships between speech and writing, is not elaborated, what he denounces as the blind prejudice of classical linguists or of common experience indeed remains a blind prejudice, on the basis of a general presupposition which is not doubt common to the accused and the prosecutor. (Derrida, 1974, p. 39)

Contratextuality of this kind is disconcerting. Backtracking, and re-working the preceding text will not help; text 11 does not predict this kind of Macro-New. And this uncertainty is precisely what Derrida is striving to achieve. Expectations are frustrated so that Derrida's own text will draw attention to itself — so that its own texture will be foregrounded. The point of this is to stop readers discarding Derrida's own text by focusing on some underlying theory or representation it seamlessly constructs. Seamful text is important for post-structuralism as a device for blocking readings which treat texts as transparent and unproblematic formulations of transcendental signifieds. Uncertainty about projection, and contradictory Macro-News are part of this textual strategy.

As with ethnomethodology, this ideological function of Derrida's discourse is clear. Its anti-rationality provides the basis for a radical critique of systems, or aspects of systems, which are taken for granted because of the effectiveness of the naturalizing discourses through which they are realized. In this respect, post-structuralism is a more attractive discourse than ethnomethodology which itself naturalizes discourses of class, gender, race and age in very uncritical ways. Post-structuralist discourse does run the danger however of having its texture co-opted by liberal humanism, which uses its rhetorical strategies to deconstruct any crit-ical theory which makes reference to both text/process and to the systems they realize.

It is for example a small step from heteroglossia (the idea that cultures are constituted by a number of conflicting discourses — e.g., of class, race, gender and age) to pluralism (the idea that people are different in individual ways). It is another small step from dialogism (the idea that conflicting discourses of class, race, gender and age manifest themselves in all texts, not just those spoken by one or another groups) to freedom of expression (the idea that everyone should be allowed to express their point of view). It is yet another small step from nego-tiation (the idea that power is redistributed through social conflict across discourses in diglossic texts) to multiculturalism (the idea that room has to be made for different people to express their different ideas in a democratic society). It does not take long in other words for Americans to arrive at anti-systemic positions such as that articulated in text 13 below where post-structuralism's concern with challenging the way in which systems are naturalized (see Derrida, 1979, on formalization and Marxist analyses of discursive practices) is taken up as challenging the existence of these systems in the first place.

WHOLE LANGUAGE

[13] We sometimes think that socio-economic status, restricted and elaborated codes, class, role, genre, well-formedness, I.Q., developmen-tal stages, and other constructs are simply blocking variables hallucinated by neo-behaviourists in linguistics, psychology, and sociology so that they would have something to run their statistical data against. They should have no a priori reality in a new theory of literacy. We all play many roles even in the same context. It depends on the mind of the beholder. Social class is a state of mind. Social class is more an attitude than a fixed state of being. Even yuppiehood can be outgrown. (Harste and Short, 1988)

When anti-rationality is co-opted to shore up progressive education in this way, the results are potentially damaging. Text 13 is part of a concerted critique of genre-based literacy programmes mounted by 'whole language' educators from Britain and America visiting Australia for the meetings of the International Read-ing Association in Brisbane in July, 1988. The basic thrust of the critique was that as a model of social practices, genre theory denies individuality, that because of this individuality social practices must be learned, not taught, and that in any case even if genres could be taught, learning hegemonic social practices such as exposi-tory writing would be depowering, not empowering, for working-class, migrant and Aboriginal children. To the extent that discourses of class, race, gender and

Table 11.6: *Synoptic Overview of Key Meanings in the Pedagogic Discourses of Science and History*

	SCIENCE	HISTORY
GRAMMAR	identifying defining relational	attributive classifying relational
DISCOURSE SEMANTICS	external conjunction (congruent)	internal & incongruent external conjunction
INTERACTION PATTERNS	nominalisation & definitions [distill]	nominalisation & macro/ hyper Theme/New [scaffold]
REGISTER	taxonomy & implication sequence constructing (*field* oriented)	text constructing; borrowed technicality (*mode* oriented)
GENRE	report: taxonomising [[explanation]] [[experiment]] 'analyse'	(generalised recount) [[report: generalising]] [[exposition]] 'interpret'

age can be denied as having any real bearing on educational questions in the first place, so much the better for this critique. What is really important is to let everyone be free to be themselves (for a response to this romanticized humanism see Martin, Christie and Rothery, 1987).

In summary, the real function of the anti-technical and anti-rational discourses surveyed here is an ideological one, which means they must be evaluated on political grounds. The point of including them here has been to highlight the way in which the pedagogic discourses of science and history are constructed for young apprentices in junior-secondary school. These students need to learn to deconstruct both technical and anti-technical discourses, and abstract and anti-rational ones to succeed in tertiary education in the late twentieth century. It makes sense to begin with technicality and abstraction, rather than the anti-discourses, which can only really be interpreted in relation to these. This is an important consideration wherever educators are concerned with introducing modern critical theory into the curriculum.

Texture in Science and Humanities

There is no easy way to summarize the configurative rapports constituting the pedagogic discourses of science and history in Australian junior-secondary schools. Table 11.6 glosses over some of the key variables, seen from the perspective of lexicogrammar, discourse semantics, interaction patterns between these two linguistic strata, register, and genre.

Grammatically, science foregrounds identifying relational processes, used to define technical terms; history relies more on attributive relational processes to assign participants to familiar classes. Semantically, science is more likely to realize logical connections between clauses and sentences than is history, which prefers to bury reasoning inside the clause. The difference between these relational process and conjunction patterns means that grammatical metaphor plays a different role in mediating between grammar and semantics in the two discourses. In science, nominalization is strongly associated with definitions; its function is to accumulate meanings so that a technical term can be defined—in science grammatical metaphor **distills**. In history on the other hand nominalization is strongly associated with realizing events as participants so that logical connections can be realized inside the clause; and at the same time nominalization is deployed to construct layers of hyper- and macro-thematic and information structure in a text—in history grammatical metaphor **scaffolds**.

Seen from the point of view of the register, what these linguistic patterns represent is that fact that science is concerned with constructing taxonomies and implication sequences while history is concerned with constructing text. In a sense the texture of scientific discourse is oriented to field, while that of history is focused on mode. One effect of this is that the 'knowledge' constructed by science is more transcendent than that developed by history. Scientific taxonomies and implication sequences tend to function in their discipline as system, not process; they are constructed as part of the semiotic potential of the field. Historical generalizations and explanations on the other hand tend to function in their discipline as text, not system; they participate in their fields as documents, to which historians must turn to find out what history means.

As far as genre is concerned, scientific discourse in junior-secondary school textbooks is organized as one large report, within which explanation and experiment genres are embedded. In contrast, history textbooks at this level are organized as long generalized recounts, within which reports and more occasionally expositions are embedded. Generically then science is about what the world is like, whereas history is about what happened. The semiosis at all levels constructs both disciplines as truth, or at best as hypothesis about what is and what happened that can be proved or disproved. The idea that the discourses of science and history are constitutive of those disciplines and negotiable, could not be more hidden.

Summing up these differences, the metaphors of analysis and synthesis are not inappropriate. Science pulls the present apart, history puts the past together. The products of analysis are the taxonomies and implication sequences constituting scientific fields while the products of synthesis are historical texts. This at least is the way the disciplines are constructed for young apprentices in Australian secondary school. As with their semogenesis in the culture, it is written discourse which is the key to their technology. As presently taught, it is a key which few students will learn to turn, with access strongly mediated by gender, race, class and age.

References

ALLPORT, G. and ALLPORT, C. (1980) *Working Lives*, Sydney, Methuen.

BAIN, M. (1979) 'At the Interfaces: the implications of opposing views of reality', Unpublished MA Thesis, Monash University.

BAKHTIN, M.M. (1986) 'The problem of speech genres', in BAKHTIN, M.M., *Speech Genres and other Late Essays* (translated by McGEE, Y.), Austin, University of Texas Press, pp. 60–102.

BARCAN, A., BLUNDEN, T., DWIGHT, A. and SHORTIS, S. (1972) *Before Yesterday*, Melbourne, Macmillan.

BAZERMAN, C. (1988) *Shaping Written Knowledge: the genre and activity of the experimental article in science*, Madison, Wisconsin, University of Wisconsin Press (Rhetoric of the Human Sciences).

BEACHAM, J. and THORPE, C. (1980) *Dear Teacher*, Canberra, Schools Commission (Schools Commission Issues Paper 6).

BENSON, L. (1957) *Plant Classification*, Boston, Heath.

BERNSTEIN, B. (1971a) *Class, Codes and Control 1: theoretical studies towards a sociology of language*, (Primary Socialization, Language and Education). [republished, with an added Appendix, by Palladin, 1974], London, Routledge and Kegan Paul.

BERNSTEIN, B. (1971b) 'Social class, language and socialisation' in ABRAMSON, A.S. *et al.* (Eds) *Current Trends in Linguistics*, **12**, The Hague, Mouton (reprinted in Bernstein 1971a/1974, pp. 193–213).

BERNSTEIN, B. (1973) *Class, Codes and Control 2: applied studies towards a sociology of language* (Primary Socialisation, Language and Education), London, Routledge and Kegan Paul.

BERNSTEIN, B. (1975) *Class, Codes and Control 3: towards a theory of educational transmissions* (Primary Socialisation, Language and Education), London, Routledge and Kegan Paul.

BERNSTEIN, B. (1984) 'Codes, modalities and cultural reproduction', in APPLE, M.W. (Ed.) *Cultural and Economic Reproduction in Education: essays on class, ideology and state*, London, Routledge and Kegan Paul, pp. 304–55.

BERNSTEIN, B. (1986) 'On pedagogic discourse', in RICHARDSON, J.G. (Ed.) *Handbook for Theory and Research in the Sociology of Education*, New York, Greenwood Press.

BERNSTEIN, B. (1987) 'Elaborated and restricted codes: an overview 1958–85', in AMMON, U., MATTHIER, K. and DITTMAR, N. (Eds) *Sociolinguistics/Soziolinguistik*, Berlin, De Gruyter.

BERNSTEIN, B. (1988) 'A sociology of the pedagogic contexts, in GEROT, L., *et al.* (Eds), pp. 13–24.

BERNSTEIN, B. (1990) *Class, Codes and Control 4: the structuring of pedagogic discourse*, London, Routledge.

BOHM, D. (1980) *Wholeness and the Implicate Order*, London, Routledge and Kegan Paul (republished by Ark paperbacks, 1983).

BRIGGS, J.P. and PEAT F.D. (1985) *Looking Glass Universe: the emerging science of wholeness*, Glasgow, Fontana.

BRIGHT, T. (1588) *Characterie*, London.

BRITTON, J., BARRS M. and BURGESS, T. (1979) 'No, no, Jeanette!', *Language in Education*, 1, University of Exeter, pp. 23–41.

BRITTON, J., BURGESS, T., MARTIN, N., MCLEOD, A. and ROSEN, H. (1975) *The Development of Writing Abilities*, London, McMillan.

BUTT, D. (1984) *The Relationship between Theme and Lexicogrammar in the Poetry of Wallace Stevens*, Sydney, Macquarie University PhD Thesis.

CASSIRER, E. (1961) *The Philosophy of Symbolic Forms*, Vol. 1, *language*, New Haven, Yale University Press.

CHAUCER, G. (1391) *A Treatise on the Astrolabe* (Early English Text Society Edition).

CHRISTIE, F. (1986) 'Learning to write: where do written texts come from?', Paper presented at the Twelfth Annual Conference of the Australian Reading Association on Text and Context, Perth.

CHRISTIE, F. (1987) 'Language and literacy: making explicit what's involved', Paper presented at the Thirteenth Annual Conference of the Australian Reading Association on Language and Learning, Sydney.

CHRISTIE, F. (1988) 'The construction of knowledge in the primary school', in GEROT, L. *et al.* (Eds), pp. 97–142.

CHRISTIE, F. (1989) 'Genres in Writing', in CHRISTIE, F. (Ed.) *Writing in Schools: study guide* Geelong, Victoria: Deakin University Press, pp. 3–48.

CHRISTIE, F. (Ed.) (1991) *Literacy in Social Processes: papers from the Inaugural Australian Systemic Network Conference, Deakin University, January 1990*, Darwin, Centre for Studies of Language in Education, Northern Territory University.

CHRISTIE, F., MARTIN, J.R. and ROTHERY, J. (1989) 'Genres make meaning: another reply to Sawyer and Watson', *English in Australia*, 90, pp. 43–59.

CHRISTIE, F. and ROTHERY, J. (1989) 'Exploring the written mode and the range of factual genres', in CHRISTIE, F. (Ed.) *Writing in Schools*, Geelong, Victoria, Deakin University Press (B. Ed. Course Study Guide), pp. 49–90.

CLYNE, M. (1991) 'Directionality, rhythm and Cultural Values in discourse', in CHRISTIE, F. (Ed.), pp. 45–59.

COLLERSON, J. (1984) 'Discovering what writing is for', in CHRISTIE, F. (Ed.) *Children Writing: reader*, Geelong, Victoria, Deakin University Press (ECT Language Studies: children writing), pp. 46–53.

CRANNY-FRANCIS, A., LEE, A., MCCORMACK, R. and MARTIN, J.R. (1991) 'Danger — shark: assessment and evaluation of a student text', in CHRISTIE, F. (Ed.), pp. 245–85.

CULL, R. and COMINO, G. (1987) *Science for Living*, 1, Milton, Queensland, Jacaranda.

CULLER, J. (1977) *Saussure*, Glasgow, Fontana (Fontana Modern Masters).

DALTON, J. (1827) *A New System of Chemical Philosophy*, London, George Wilson.

DANES, F. (1974) 'Functional sentence perspective and the organization of the text', in DANES, F. (Ed.) *Papers on functional sentence perspective*, The Hague, Mouton, pp. 106–28.

DARWIN, C. (1979) *The Origin of the Species by means of Natural Selection* (with a new preface by Patricia Horan) New York, Avenel Books (original 1859).

DEPARTMENT of EDUCATION, VICTORIA (1981) *Science in the Primary School: providing for enquiry*, Vol. 1, Melbourne, Department of Education.

DEREWIANKA, B. (1989) *Exploring how Text Works*, Sydney, Primary English Teaching Association.

DEREWIANKA, B. (1990) 'Rocks in the head: children and the language of geology', in CARTER, R. (Ed.) *Knowledge about Language and the Curriculum: the LINC reader*, London, Hodder and Stoughton, pp. 197–215.

DERRIDA, J. (1974) *Of Grammatology* (translated by SPIVAK, G.C.), Baltimore, John Hopkins University Press.

DERRIDA, J. (1987) 'Some questions and responses', in FABB, N., ATTRIDGE, D., DURANT, A. and MACCABE, C. (Eds) *The Linguistics of Writing: arguments between language and literature*, Manchester, Manchester University Press, pp. 252–64.

DIJKSTERHUIS, E.J. (1961) *The Mechanization of the World Picture*, London, Oxford (republished by Princeton University Press, 1986).

DISADVANTAGED SCHOOLS PROGRAM (1986a) 'Equality of Outcomes: an issue paper', Sydney Disadvantaged Schools Program, NSW Department of Education.

DISADVANTAGED SCHOOLS PROGRAM (1986b) 'What is the DSP?', Sydney Disadvantaged Schools Program, NSW Department of Education.

DISADVANTAGED SCHOOLS PROGRAM (1988) *Teaching Factual Writing: a genre based approach*, Sydney, Disadvantaged Schools Program, Metropolitan East Region[1].

DISADVANTAGED SCHOOLS PROGRAM (1989a) *The Report Genre*, Sydney Disadvantaged Schools Program, Metropolitan East Region.

DISADVANTAGED SCHOOLS PROGRAM (1989b) *The Discussion Genre*, Sydney Disadvantaged Schools Program, Metropolitan East Region.

DISADVANTAGED SCHOOLS PROGRAM (1990) *Assessing Writing: scientific reports*, Sydney Disadvantaged Schools Program, Metropolitan East Region.

DISADVANTAGED SCHOOLS PROGRAM (1991) *The Recount Genre*, Sydney Disadvantaged Schools Program, Metropolitan East Region.

EGGINS, S. (1986) 'Technicality, grammatical metaphor and the writings of the ethnomethodologists', Department of Linguistics, University of Sydney (mimeo).

EGGINS, S., WIGNELL, P. and MARTIN, V.R. (1987) 'The discourse of history: distancing the recoverable past', *Writing Project: report 1987* (Working Papers in Linguistics 5), Department of Linguistics, University of Sydney, pp. 25–65 (to be republished in GHADESSY, M. (Ed.) *Register Analysis: theory into practice*, London, Pinter).

---

[1] Available from Met East DSP, Erskineville Public School, corner Bridge and Swanson streets, Erskineville, NSW.

FAIRCLOUGH, N. (1988) 'Register, power and socio-semantic change', in BIRCH, D. and O'TOOLE, M. (Eds) *The Functions of Style*, London, Pinter, pp. 111–25.

FRICKE, J. (1988) 'Aerogels', *Scientific American* 258.5. May 1988, pp. 92–7.

FRIERE, P. (1974) *Education for Critical Consciousness*, London, Sheed and Ward.

FRIES, P.H. (1981) 'On the status of theme in English: arguments from discourse', *Forum Linguisticum*, **6**, **1**, pp. 1–38. (republished in PETOFI, J.S. and SOZER, E. (Eds) (1983) *Micro and macro Connexity of Texts*, Hamburg, Helmut Buske Verlag (Papers in Textlinguistics 45), pp. 116–52).

FRIES, P.H. (forthcoming) 'Towards a discussion of the flow of information in written text', in Gregory, M. (Ed.) *Relations and Functions within and around Text*, Amsterdam, Benjamins.

GEROT, L., OLDENBURG, J. and VAN LEEUWEN, T. (Eds) (1988) *Language and Socialisation, Home and School: proceedings from the Working Conference on Language in Education, Macquarie University, 17–21 November 1986*, Macquarie University.

GIBLETT, R. and O'CARROLL, J. (Eds) (1990) *Discipline — Dialogue — Difference: proceedings of the Language in Education Conference, Murdoch University, December 1989*, Perth, 4D Duration Publications, School of Humanities, Murdoch University.

GOULD, S.J. (1989) *Wonderful Life: the Burgess Shale and the nature of history*, London, Hutchinson Radius (republished by Penguin, 1991).

GRAHAM, B. (1986) 'Learning the language of schooling: considerations and concerns in the Aboriginal context', *Keynote Address Book*, Australia and New Zealand Conference on the First Years of School, Sydney, 12–15 May.

GRAY, B. (1987), 'How natural is 'natural' language teaching: employing wholistic methodology in the classroom', *Australian Journal of Early Childhood*, **12**, 4, pp. 3–19.

GREGORY, M. (1967) 'Aspects of varieties differentiation', *Journal of Linguistics*, **3**, pp. 177–98.

GRIBBIN, J. (1985) *In Search of Schrödinger's Cat: quantum physics and reality*, London, Corgi Books.

HABERMAS, J. (1984) *The Theory of Communicative Action: reason and the rationalization of society* (MCCARTHY, T. trans.), **1**, Boston, Beacon Press.

HAGÈGE, C. (1988) *Leçon Inaugurale*, Paris, Collège de France.

HALL, R.A. (1987) 'Deconstructing Derrida on language', *Linguistics and Pseudo-Linguistics: selected essays 1965–1985*, Amsterdam, John Benjamins, pp. 116–21.

HALLIDAY, M.A.K. (1967) *Grammar, Society and the Noun*, London, H.K. Lewis, for University College London (republished in HALLIDAY, M.A.K. (1977) *Aims and Perspectives in Linguistics*. Applied Linguistics Association of Australia, (Occasional Papers 1), pp. 1–18.)

HALLIDAY, M.A.K. (1975) *Learning How to Mean: explorations in the development of language*, London, Edward Arnold.

HALLIDAY, M.A.K. (1978) *Language as Social Semiotic: the social interpretation of language and meaning*, London, Edward Arnold.

HALLIDAY, M.A.K. (1979) 'Modes of meaning and modes of expression: types of grammatical structure, and their determination by different semantic functions', in ALLERTON, D.J., CARNEY, E. and HOLDCROFT, D. (Eds) *Function and Context in Linguistic Analysis: essays offered to William Haas*, Cambridge, Cambridge University Press, pp. 57–79.

References

HALLIDAY, M.A.K. (1981) 'Text semantics and clause grammar: some patterns of realization', in COPELAND, J.E. and DAVIS, P.W. (Eds) *The Seventh LACUS Forum*, Columbia, S.C. Hornbeam Press, pp. 31–59.

HALLIDAY, M.A.K. (1982) 'How is a text like a clause?', in Allen, S. (Ed.) *Text Processing: text analysis and generation, text typology and attribution*, Stockholm, Almquist and Wiksell International, pp. 209–47.

HALLIDAY, M.A.K. (1985a) *An Introduction to Functional Grammar*, London, Edward Arnold.

HALLIDAY, M.A.K. (1985b) 'Context of situation', in HALLIDAY, M.A.K. and HASAN, R. (Eds) *Language, Context and Text*, Geelong, Victoria, Deakin University Press, pp. 3–14. (republished by Oxford University Press, 1989).

HALLIDAY, M.A.K. (1985c) *Spoken and Written Language*, Geelong, Victoria, Deakin University Press (republished by Oxford University Press, 1989).

HALLIDAY, M.A.K. (1986) 'Spoken and written modes of meaning', in HOROWITZ, R. and SAMUELS, S.J. (Eds) *Comprehending Oral and Written Language*, New York, Academic Press, pp. 55–82.

HALLIDAY, M.A.K. (1987) 'Poetry as scientific discourse: the nuclear sections of Tennyson's *In Memoriam*', in BIRCH, D. and O'TOOLE, M. (Eds) *The Functions of Style*, London, Pinter, pp. 31–44.

HALLIDAY, M.A.K. (1990) 'New ways of meaning: a challenge to applied linguistics', *Journal of Applied Linguistics* (Greek Applied Linguistics Association), **6**, pp. 7–36. (Ninth World Congress of Applied Linguistics Special Issue). (reprinted in PÜTZ, M. (Ed.), *Thirty Years of Linguistic Evolution*, Amsterdam, Benjamins, 1992)

HALLIDAY, M.A.K. (1992) 'Systemic grammar and the concept of a "science of language"', *Waiguoyu* (*Foreign Languages*) (Shanghai International Studies University), **92**, 2, pp. 1–9 (reprinted in Network, **19**, pp. 55–64).

HALLIDAY, M.A.K. and HASAN, R. (1976) *Cohesion in English*, London, Longman (English Language Series 9).

HALLIDAY, M.A.K. and HASAN, R. (1980) *Text and Context: aspects of language in a social-semiotic perspective*, Sophia Linguistica VI, Tokyo: The Graduate School of Languages and Linguistics & the Linguistic Institute for International Communication, Sophia University (new edition published as HALLIDAY, M.A.K. and HASAN, R. (1985) *Language, context and text: aspects of language in a social-semiotic perspective*, Geelong, Victoria, Deakin University Press (republished by Oxford University Press, 1989)).

HALLIDAY, M.A.K., MCINTOSH, A. and STREVENS, P. (1964) *The Linguistic Sciences and Language Teaching*, London, Longman (Longman's Linguistics Library).

HALLIDAY, M.A.K. and MATTHIESSEN, C.M. (forthcoming) *Construing Experience through Meaning: a language-based approach to cognition*.

HAMILTON, J.H. and MARUHN, J.A. (1986) 'Exotic atomic nuclei', *Scientific American*, July, pp. 74–83.

HAMMOND, J. (1987) 'An overview of the genre-based approach to the teaching of writing in Australia', *Australian Review of Applied Linguistics*, **10**, 2, pp. 163–81.

HAMMOND, J. (1990) 'Is learning to read and write the same as learning to speak?', in CHRISTIE, F. (Ed.) *Literacy for a Changing World*, Melbourne: Australian Council for Educational Research (A Fresh Look at the Basics), pp. 26–53.

HARSTE, J.C. and SHORT, K.G. (1988) 'What educational difference does your theory of language make?', Paper presented at the Language in Learning Symposium, Mount Gravatt, Queensland, Australia.

HASAN, R. (1973) 'Code, register and social dialect', Appendix to BERNSTEIN, B. *Class, Codes and Control 2: applied studies towards a sociology of language*, London, Routledge and Kegan Paul (Primary Socialization, Language and Education), pp. 253–92.

HASAN, R. (1985) 'The texture of a text', in HALLIDAY, M.A.K. and HASAN, R. *Language, Context and Text*, Geelong, Victoria, Deakin University Press, pp. 70–96 (republished by Oxford University Press, 1989).

HASAN, R. (1986) 'The ontogenesis of ideology: an interpretation of mother child talk', in THREADGOLD, T. GROSZ, E.A., KRESS, G. and HALLIDAY, M.A.K. (Eds) *Language, Semiotics, Ideology*, Sydney, Sydney Association for Studies in Society and Culture (Sydney Studies in Society and Culture 3), pp. 125–46.

HASAN, R. (1987a) 'The grammarian's dream: lexis as most delicate grammar', in HALLIDAY, M.A.K. and FAWCETT, R.P. (Eds) *New Developments in Systemic Linguistics 1: theory and description*, London, Pinter, pp. 184–211.

HASAN, R. (1987b) 'Directions from structuralism', in FABB, N., ATTRIDGE, D., Durnat, A. and MACCABE, C. (Eds) *The Linguistics of Writing: arguments between language and literature*, Manchester, Manchester University Press, pp. 103–22.

HASAN, R. (1990) 'Semantic variation and sociolinguistics', *Australian Journal of Linguistics*, **9**, 2, pp. 221–76.

HASAN, R. (forthcoming) 'Offers in the making: a systemic-functional approach', School of English and Linguistics, Macquarie University (mimeo).

HEADING, K.E.G., PROVIS, D.F., SCOTT, T.D., SMITH, J.E. and SMITH, R.T. (1967) *Science for Secondary Schools*, **2**, Adelaide, Rigby.

HEATH, T. (1913) *Aristarchus of Samos, the ancient Copernicus*, Oxford, Clarendon (republished by Dover Publications 1981).

HEFFERNAN, D.A. and LEARMONTH, M.S. (1981) *The World of Science — Book 2*, Melbourne, Longman Cheshire.

HEFFERNAN, D.A. and LEARMONTH, M.S. (1982) *The World of Science — Book 3*, Melbourne, Longman Cheshire.

HEFFERNAN, D.A. and LEARMONTH M.S. (1983) *The World of Science — Book 4*, Melbourne, Longman Cheshire.

HEISENBERG, W. (1959) *Physics and Philosophy: the revolution in modern science*, New York, Harper and Row.

HELLIGE, J.B. (1978) 'Visual laterality patterns for pure- versus mixed- list presentation', *Journal of Experimental Psychology*, **4**, **1**.

HILL, G.C. and HOLMAN, J.S. (1978) *Chemistry in Context*, Walton on Thames, Thomans Nelson, (2nd ed. 1983).

HORVATH, B. (1985) *Variation in Australian English: the sociolects of Sydney*, Cambridge, (Cambridge Studies in Linguistics 45), Cambridge University Press.

HUDDLESTON, R.D., HUDSON, R.A., WINTER, E.O. and HENRICI, A. (1968) *Sentence and Clause in Scientific English*, London, University College London, Communication Research Centre. (Report of O.S.T.I. Project 'The Linguistic Properties of Scientific English')

INTERMEDIATE SCIENCE CURRICULUM STUDY (1976) *Well-being: probing the natural world*, Hong Kong, Martin Educational.

References

JOHNSON, A.W. and EARLE, T. (1987) *The Evolution of Human Societies: from foraging group to agrarian state*, Stanford, CA, Stanford University Press.

JUNIOR SECONDARY SCIENCE PROJECT (1968) *When Substances are Mixed*, Melbourne, Longman, Cheshire (2nd ed. 1973).

KALANTZIS, M. and COPE, B. (1984) 'Multiculturalism and education policy', in G. BOTTOMLEY (Ed.), *Ethnicity, Class and Gender*, London, George Allen and Unwin.

KALANTZIS, M. and WIGNELL, P. (1988) *Explain? Argue? Discuss?: writing for essays and exams*, Sydney, Common Ground.

KING, P. (in prep.) *Spoken and written science and engineering text* (Prepublication draft), Birmingham, English Department, University of Birmingham.

KRESS, G. (1988a) 'Textual matters: the social effectiveness of style', in BIRCH, D. and O'TOOLE, M. (Eds) *The Functions of Style*, London, Pinter, pp. 126–41.

KRESS, G. (1988b) 'Barely the basics? New directions in writing', *Education Australia*, Term 1, pp. 12–15.

KUHN, T.S. (1962) *The Structure of Scientific Revolutions*, Chicago, University of Chicago Press.

LEE, A. and GREEN, B. (1990) 'Staging the differences: on school literacy and the socially-critical curriculum', in GIBLETT, R. and O'CARROLL, J. (Eds), pp. 225–61.

LEE, B. (1985) 'Peirce, Frege, Saussure and Whorf: the semiotic mediation of ontology', in MERTZ, E. and PARMENTIER, R.J. (Eds) *Semiotic mediation: socio-cultural and psychological perspectives*, Orlando, Academic Press, pp. 100–28.

LEMKE, J.L. (1982) 'Talking physics', *Physics Education*, **17**, pp. 262–7.

LEMKE, J.L. (1983) *Classroom Communication of Science*, Final Report to the US National Science Foundation (ERIC Document Reproduction Service No. ED 222 346).

LEMKE, J.L. (1984) *Semiotics and Education*, Toronto, Toronto Semiotic Circle (Monographs, Working Papers and Publications 2).

LEMKE, J.L. (1985) *Using Language in the Classroom*, Geelong, Vic, Deakin University Press (republished by Oxford University Press, 1989).

LEMKE, J.L. (1988) 'Genres, semantics and classroom education', *Linguistics and Education*, **1**, 1, pp. 81–100.

LEMKE, J.L. (1990a) *Talking Science: language, learning and values*, Norwood, NJ, Ablex (Language and Educational Processes).

LEMKE, J.L. (1990b) 'Technical discourse and technocratic ideology', in HALLIDAY, M.A.K., GIBBONS, J. and NICHOLAS, H. (Eds) *Learning, Keeping and Using Language: selected papers from the 8th World Congress of Applied Linguistics*, **2**, Amsterdam, Benjamin, pp. 435–60.

LEMKE, J.L. (1992) 'New challenges for systemic-functional linguistics: dialect diversity and language change', *Network*, **18**, pp. 61–68.

LEVINSON, S.C. (1983) *Pragmatics*, Cambridge, Cambridge University Press (Cambridge Textbooks in Linguistics).

LINNAEUS, C. (1957) *Species Plantarum*, London, Royal Society (Facsimile of 1753 edition).

LITTLEFAIR, A. (1991) *Reading all Types of Writing: the importance of genre and register for reading development*, Milton Keynes, Open University Press (Rethinking Reading).

Lucy, J.A. (1985) 'Whorf's view of the linguistic mediation of thought', in Mertz, E. and Parmentier, R.J. (Eds) *Semiotic mediation: sociocultural and psychological perspectives*, Orlando, Academic Press, pp. 74–97.

Lucy, J.A. and Wertsch, J.V. (1987) 'Vygotsky and Whorf: a comparative analysis', in Hickman, M. (Ed.) *Social and Functional Approaches to Language and Thought*, Orlando, Academic Press, pp. 67–86.

Macken, M., Martin, J.R., Kress, G., Kalantzis, M., Rothery, J. and Cope, W. (1989a) *An Approach to Writing K-12, Introduction,* 1 Sydney, Literacy and Education Research Network and Directorate of Studies, NSW Department of Education[2].

Macken, M., Martin, J.R., Kress, G., Kalantzis, M., Rothery, J. and Cope, W. (1989b) *An Approach to Writing K-12: Factual Writing: a teaching unit based on reports about sea mammals,* 2, Sydney, Literacy and Education Research Network and Directorate of Studies, NSW Department of Education.

Macken, M., Martin, J.R., Kress, G., Kalantzis, M., Rothery, J. and Cope, W. (1989c) *An Approach to Writing K-12: Story Writing: a teaching unit based on narratives, news stories and fairy tales,* 3, Sydney, Literacy and Education Research Network and Directorate of Studies, NSW Department of Education.

Macken, M., Martin, J.R., Kress, G., Kalantzis, M., Rothery, J. and Cope, W. (1989d) *An approach to Writing K-12: The Theory and Practice of Genre-Based Writing,* 4, Sydney, Literacy and Education Research Network and Directorate of Studies, NSW Department of Education.

Maclean, R. (1984) 'Expository writing', in Christie, F. (Ed.) *Children Writing: study guide*, Geelong, Victoria, Deakin University Press (ECT Language Studies: children writing), pp. 159–80.

Malinowski, B. (1935) *Coral Gardens and their Magic,* 2, London, Allen and Unwin.

Mann, W.C. and Thompson S. (Eds) 1992. *Discourse Description: diverse analyses of a fund raising text.* Amsterdam: Benjamins (Pragmatics and Beyond: new series).

Martin, J.R. (1983) 'Conjunction: the logic of English text', in Petöfi, J.S. and Sözer, E. (Eds) *Micro and Macro Connexity of Texts*, Hamburg, Helmut Buske (Papers in Textlinguistics 45), pp. 1–72.

Martin, J.R. (1984a) 'Language, register and genre', in Christie, F. (Ed.) *Children Writing: reader*, Geelong, Victoria, Deakin University Press (ECT 418 Language Studies), pp. 21–30.

Martin, J.R (1984b) 'Systemic functional linguistics and an understanding of written text', in Bartlett, B. and Carr, J. (Eds) *Language in Education Workshop: a report of proceedings*, Brisbane, Centre for Research and Learning in Literacy, Brisbane CAE, Mr Gravatt Campus, pp. 22–40. (republished in Martin, J.R. and Rothery, J. *Working Papers in Linguistics 5 Writing Project — report 1986*, Department of Linguistics, University of Sydney, pp. 91–110).

Martin, J.R. (1985a) 'Process and Text: two aspects of human semiosis', in Benson, J.D. and Greaves, W.S. (Eds) *Systemic Perspectives on Discourse: selected theoretical papers from the 9th International Systemic Congress,* 1 Norwood, NJ, Ablex, pp. 248–74.

[2] Available Common Ground Publishing, 6A Nelson St, Annandale NSW 2038.

MARTIN, J.R. (1985b) *Factual Writing: exploring and challenging social reality*, Geelong, Victoria, Deakin University Press (republished by Oxford University Press, 1989).

MARTIN, J.R. (1986a) 'Prewriting: oral models for written text', in WALSHE, R.D., MARCH, P. and JENSEN, D. (Eds) *Writing and Learning in Australia*, Melbourne, Dellasta Books, pp. 138–42. (unabridged version published in *Prospect: The Journal of the Adult Migrant Education Program*, **3**, **1**, 1987, pp. 75–90).

MARTIN, J.R. (1986b) 'Intervening in the process of writing development', in PAINTER, G. and MARTIN, J.R. (Eds) *Writing to Mean: teaching genres across the curriculum*, Applied Linguistics Association of Australia (Occasional Papers 9), pp. 11–43.

MARTIN, J.R. (1990) 'Language and control: fighting with words', in WALTON, C. and EGGINGTON, W. (Eds) *Language: maintenance, power and education in Australian Aboriginal contexts*, Darwin, NT, Northern Territory University Press, pp. 12–43.

MARTIN, J.R. (1991) 'Critical literacy: the role of a functional model of language', *Australian Journal of Reading*, **14**, 2 (Focus Issue — Literacy Research in Australia: selected perspectives by FREEBODY, P. and SHORTLAND-JONES, B. (Eds)), pp. 117–132.

MARTIN, J.R. (1992) *English Text: system and structure*, Amsterdam, Benjamins.

MARTIN, J.R. (forthcoming) 'Lexical cohesion, field and genre: parcelling experience and discourse goals', COPELAND, J.E. (Ed.) *Text Semantics and Discourse Semantics*, Houston, Rice University Press, in (in press) (also to be published in *Systemic Functional Linguistics Forum*].

MARTIN, J.R., CHRISTIE, F. and ROTHERY, J. (1987) 'Social processes in education', in REID, I. (Ed.) *The Place of Genre in Learning*, Geelong, Victoria, Centre for Studies in Literary Education, Deakin University (Typereader Publications 1), pp. 58–82. (fuller version published in *The Teaching of English: Journal of the English Teachers Association of New South Wales*, **53**, (1987), pp. 3–22).

MARTIN, J.R. and PETERS, P. (1985) 'On the analysis of exposition', in HASAN, R. (Ed.) *Discourse on Discourse: workshop reports from the Macquarie Workshop on Discourse Analysis*, Applied Linguistics Association of Australia (Occasional Papers 7), pp. 61–92.

MARTIN, J.R. and ROTHERY, J. (1986) *Writing Project Report* (Working Papers in Linguistics 4), Sydney, Department of Linguistics, University of Sydney.

MARTIN, J.R., WIGNELL, P., EGGINS, S. and ROTHERY, J. (1988) 'Secret English: discourse technology in a junior secondary school', GEROT, L. *et al.* (Eds), pp. 143–73.

MARTINEC, R., WIGNELL, P. and MARTIN, J.R. (in progress) 'The role of documents in history textbooks', Unpublished manuscript, Department of Linguistics, University of Sydney.

MATTHIESSEN, C.M. (1988) 'Representational issues in systemic functional grammar', in BENSON, J.D. and GREAVES, W.S. (Eds) *Systemic Functional Approaches to Discourse*, Norwood, NJ, Ablex, pp. 136–75.

MATTHIESSEN, C.M. (1992) 'Language on language: the grammar of semiosis', *Social Semiotics*, **1**, 2, pp. 69–111.

MATTHIESSEN, C.M. and HALLIDAY, M.A.K. (1992) *Systemic Functional Grammar*.

MAXWELL, J.C. (1881) *An Elementary Treatise on Electricity*, Oxford, Clarendon.

McCusker, B. (1983) 'Fundamental particles', in Williams, R. (Ed.) *The Best of the Science Show*, Melbourne, Thomas Nelson (in association with the Australian Broadcasting Commission), pp. 235–42.

McMullen, A. and Williams, J.L. (1971) *On Course Mathematics*, Sydney, Australia, Macmillan.

McNamara, J. (1989) 'The writing in science and history project: the research questions and implications for teachers', in Christie, F. (Ed.) *Writing in Schools: reader*, Geelong, Victoria, Deakin University Press, pp. 24–35.

Medawar, P. (1984) *Pluto's Republic*, Oxford, Oxford University Press.

Messel, H., Crocker, R.L. and Barker, E.N. (1964) *Science for High School Students*, Sydney, Nuclear Research Foundation, University of Sydney.

Michalske, T.A. and Bunker, B.C. (1987) 'The fracturing of glass', *Scientific American*, December.

Moore, W.G. (1966) *A Dictionary of Geography*, London, Penguin Books (Original work published in 1949).

Morris, A. and Stewart-Dore, N. (1984) *Learning to Read from Text: effective reading in the content areas*, Sydney, Addison-Wesley.

Nesbitt, C. and Plum, G. (1988) 'Probabilities in a systemic-functional grammar: the clause complex in English', in Fawcett, R.P. and Young, D. (Eds) *New Developments in Systemic Linguistics 2: theory and application*, London, Pinter, pp. 6–38.

Newkirk, T. (1984) 'Archimedes' dream', *Language Arts*, **61**, 4, pp. 341–50.

Newton, I. (1704) *Opticks, or a Treatise of the Reflections Refractions Inflections and Colours of Light*, New York, Dover Publications 1952 (London, G. Bell and Sons, 1931; based on the Fourth Edition, London 1730; originally published 1704).

NSW Department of Education (1984) *Science 7–10 Syllabus*, Sydney, NSW Department of Education.

NSW Department of Education (1987) *Writing K-12*, Sydney, NSW Department of Education.

Oldroyd, D. (1986) *The Arch of Knowledge: an introductory study of the history of the philosophy and methodology of science*, Kensington, NSW, New South Wales University Press.

Osaka Port and Harbour Bureau (n.d.) (circa 1987) *Technoport Osaka: a centennial project of the Municipality of Osaka*, Osaka, City of Osaka Port and Harbour Bureau.

Painter, C. (1984) *Into the Mother Tongue: a case study of early language development*, London, Pinter.

Painter, C. (1985) *Learning the Mother Tongue*, Geelong, Victoria, Deakin University Press (republished by Oxford University Press, 1989).

Painter, C. (1986) 'The role of interaction in learning to speak and learning to write', in Painter, C. and Martin, J.R. (Eds) *Writing to Mean: teaching genres across the curriculum*, Applied Linguistics Association of Australia (Occasional Papers 9), pp. 62–97.

Painter, C. and Martin, J.R. (Eds) (1986) *Writing to Mean: teaching genres across the curriculum*, Applied Linguistics Association of Australia (Occasional Papers 9).

Parkes, A.A., Couchman, K.E. and Jones, S.B. (1978) *Betty and Jim: year six mathematics*, Sydney, Shakespeare Head Press.

References

PENROSE, R. (1989) *The Emperor's New Mind: concerning computers, minds, and the laws of physics*, London, Oxford University Press (republished by Vintage Books, 1990).

PRIESTLEY, J. (1767) *The History and Present State of Electricity*, London.

PRIGOGINE, I. and STENGERS, I. (1984) *Order out of Chaos: man's new dialogue with nature*, London, William Heinemann (Republished by Fontana paperbacks, 1985).

RAVELLI, L. (1985) *Metaphor, Mode and Complexity: a exploration of co-varying patterns*, BA Hons Thesis, Department of Linguistics, University of Sydney.

REID, I. (1987)[3] *The Place of Genre in Learning: current debates*, Geelong, Victoria, Centre for Studies in Literary Education, Deakin University (Typereader Publications 1).

REID, T.B.W. (1956) 'Linguistics, structuralism, philology', *Archivum Linguisticum*, **8**, pp. 28–37.

RIST, J.M. (1978) *The Stoics*, Berkeley, University of California Press (esp. Chapter 3, Andreas Graeser 'The Stoic theory of meaning', pp. 77–100).

ROTHERY, J. (1986) 'Teaching writing in the primary school: a genre-based approach to the development of writing abilities', *Writing Project Report* 1986 (Working Papers in Linguistics 4) Sydney, Department of Linguistics, University of Sydney, pp. 3–62).

ROTHERY, J. (1989a) 'Exploring the written mode and the range of factual genres', in CHRISTIE, F. (Ed.), *Writing in Schools: study guide*, Geelong, Victoria, Deakin University Press, pp. 49–90.

ROTHERY, J. (1989b) 'Learning about language', in HASAN, R. and MARTIN, J.R. (Eds) *Language Development: learning language, learning culture*, Norwood, NJ., Ablex (Advances in Discourse Processes 27 — Meaning and Choice in Language: studies for Michael Halliday), pp. 199–256.

SACHS, H., SCHEGLOFF, E. and JEFFERSON, G. (1974) 'A simplest systematics for the organization of turn-taking in conversation', *Language*, **50**, **4**, pp. 696–735.

SALE, C., FRIEDMAN, B. and WILSON, G. (1980) *Our Changing World. Book I: The Vanishing Natural Ecosystem*, Melbourne, Longman Cheshire.

SALMON, V. (1966) 'Language planning in seventeenth century England: its context and aims', in BAZELL, C.E., CATFORD, J.C., HALLIDAY, M.A.K. and ROBINS, R.H. *In Memory of JR Firth*, London, Longmans (Longmans Linguistics Library), pp. 370–97.

SALMON, V. (1979) *The Study of Language in Seventeenth Century England*, Amsterdam, Benjamins.

SANKOFF, D. and LABERGE, S. (1978) 'The linguistics marketplace and the statistical explanation of variability', SANKOFF, D. (Ed.) *Linguistic Variation: models and methods*, New York, Academic Press.

SAUSSURE, F. (1966) *Course in General Linguistics*, in BALLY, C. and SECHEHAYE, A. (Eds) (Original work published in 1915), London, McGraw-Hill.

SAWYER, W. and WATSON, K. (1987) in REID, I. (Ed.) *The Place of Genre in Learning: current debates*, Geelong, Vic, Centre for Studies in Literary Education, Deakin University (Typereader Publications no. 1), pp. 46–57.

---

[3] There is in fact no date of publication on this volume; the book however first appeared in Australia in 1987.

SECONDARY SCHOOLS BOARD (1984–1985) *New South Wales School Certificate Syllabus in Geography*, Sydney, NSW Secondary Schools Board.

SHEA, N. (1988) *The Language of Junior Secondary Science Textbooks*, BA Hons Thesis, Department of Linguistics, University of Sydney.

SIMMELHAIG, H. and SPENCELEY, G.F.R. (1984) *For Australia's Sake*, Melbourne, Nelson,

SIMON, H.A. (1969) *The Sciences of the Artificial*, Cambridge, Mass, MIT Press (2nd ed. 1981).

SLATER, P. (1983) *A Field Guide to Australian Birds*, **1**, **2**, Adelaide, Rigby.

SPENCER, E. (1983) *Written Work in Scottish Secondary Schools*, Edinburgh, Scottish Council for Research in Education.

TAYLOR, C. (1979) *The English of High School Textbooks*, Canberra, Australian Government Publishing Service (Education Research and Development Committee, Report 18).

THIBAULT, P. (1987) 'An interview with Michael Halliday', in STEELE, R. and THREADGOLD, T. (Eds) *Language Topics: essays in honour of Michael Halliday*, **2**, Amsterdam, Benjamins, pp. 599–627.

THIBAULT, P. (1989) 'Genres, social action and pedagogy: towards a critical social semiotic account', *Southern Review*, **22**, 3, pp. 338–62.

THOMPSON, J.B. (1984) *Studies in the Theory of Ideology*, Cambridge, Polity.

THREADGOLD, T. (1988) 'The genre debate', *Southern Review*, **21**, 3, pp. 315–30.

THREADGOLD. T. (1989) 'Talking about genre: ideologies and incompatible discourses', *Cultural Studies*, **3**, 1, pp. 101–27.

THREADGOLD, T., GROSZ, E.A., KRESS, G. and HALLIDAY, M.A.K. (1986) (Eds) *Language, Semiotics, Ideology*, Sydney: Sydney Association for Studies in Society and Culture (Sydney Studies in Society and Culture 3).

TREVARTHEN, C. (1992) 'Editorial: natural semiotics', *The Semiotic Review of Books*, 31 January 1992 (A Publication of the Toronto Semiotic Circle), **1**, 2.

TREWARTHA, G.T. (1986) *A Introduction to Climate*, New York, McGraw Hill.

VICKERY, R.L., LAKE, J.H., MCKENNA, L.N. and RYAN, A.S. (1978) *The process Way to Science*, Melbourne, Jacaranda Press.

VICTORIAN PRIMARY SCHOOL SCIENCE SYLLABUS. (1981) Department of Education, Victoria.

WELLS, R. (1960) 'Nominal and verbal style', in SEBEOK, T.A. (Ed.) *Style in Language*, New York, pp. 213–220.

WHITE, J. (1986) 'The writing on the wall: the beginning or end of a girl's career?', *Women's Studies International Forum*, **9**, 5, pp. 561–74, (republished in CHRISTIE, F. (Ed.) *Writing in Schools: reader*, Geelong, Victoria, Deakin University Press, pp. 61–72).

WHITE, J. and WELFORD, G. (1987) *The Language of Science: making and interpreting observations*, London, Assessment of Performance Unit, Department of Education and Science,

WHORF, B.L. (1950) 'An American Indian model of the universe', *International Journal of American Linguistics*, **16**, pp. 67–72 (reprinted in CARROLL, J.B. (Ed.) (1956) *Language, Thought and Reality: selected writing of Benjamin Lee Whorf*, Cambridge, Mass, MIT Press, pp. 57–64).

WIGNELL, P. (1987) 'In your own words', *Writing Project Report 1987* (Working Papers in Linguistics 5) Department of Linguistics, University of Sydney, pp. 1–24.

References

WIGNELL, P. (1988) *The Language of Social Literacy: a linguistic analysis of the materials in action in Years 7 & 8*, Sydney, Common Ground* (Social Literacy Monograph Series 41).

WILKINS, J. (1668) *An Essay towards a Real Character, and a Philosophical Language*, London: for Sa. Gellibrand, and for John Martyn.

WINOGRAD, T. (1983) *Language as a Cognitive Process: Syntax*, 1, Reading, Mass, Addison-Wesley.

YNGVE, V.H. (1986) 'To be a scientist', *The Thirtheenth LACUS Forum*, Lake Bluff, IL, Linguistic Association of Canada and the United States, pp. 5–25.

Index

abstraction 34–35, 211–212, 226–228
acronym 29–30, 229
ambiguity 77–79, 130

biography 196

cause 61, 64, 77, 87, 90–92
Chaucer 13, 56–57, 88–89
childism 194, 255
Chinese,
   charactery 5, 108
   grammar 126–136
classification 138, 156, 170–174,
   228–233, 73–74, 126
   – processes 177–179
clause 7, 78–82, 88–93, 114, 117
clause complex 56, 101–102
common sense 6, 15, 21, 112, 116, 126,
   168–170, 204–205
complementarity 113–119
composition 73, 126, 138–139, 156,
   173–174
   – processes 179–182
congruent, *see* grammatical metaphor
CONJUNCTION 233–236
content plane 29
context 32–36
   see also field, genre, ideology, mode,
     register, tenor
cryptotypes 113–114, 117

Darwin 93–104, 121
definition 72–73, 148–152, 209–210
density, *see* lexical density
derivational morphology 12, 62
Derrida (deconstruction of) 262–265
diagrams 174–177
dialect 4, 86–87

discourse semantics 32
distillation 30, 172, 267
dynamic (perspective) 47–48, 117–118
dynamic open system 108–112

ethnomethodology (deconstruction of)
   259–262
expansion 58, 61, 114
experimental science 7, 58, 64, 81
experiments 183–186, 192–193
explanations 156–157, 179, 181–182,
   191–192, 208–209, 222–224
expositions 196–197, 214–215, 222–224
expression plane 29
external relation, *see* logical-semantic
   relations

field 32, 33

genre 35–36
   – as particle 257–258
   – as wave 257–258
Given, *see* INFORMATION, New
grammatical metaphor 13–15, 18, 30, 56,
   60, 63, 79–82, 83, 116, 119, 128,
   218–219, 226–227, 235–241
grammatics 12, 68, 103, 115
graphology 32
Greek 12, 16, 88, 116

Hjelmslev 17
humanists 4, 16
Hyper-New 247–249, 256–257
Hyper-Theme 244–247, 256–257
hyponymy 126
   see also classification
hypotaxis 57, 102, 130
   see also clause complex

ideational metafunction 29, 88, 90
IDENTIFICATION 28
ideology 10, 37, 132
impersonal style 58, 66
implication sequence 156–161, 211
indeterminacy 20, 111, 118
INFORMATION 241–242
  see also New
information flow 64, 81–82, 92, 95
instantiation 42–45
  see also system
interpersonal metafunction 29, 88, 90
intertextuality 46
intricacy, grammatical 59, 67, 117

jargon 69–70, 159–162, 205

language,
  – as construction of experience 8, 10,
    16, 119
  – as dynamic open system 15, 108–110
  – as expression/instrument 4
  – as reflection 8
  – as ritual 68, 84
  – as social semiotic 18, 110, 118
  – as system-&-process 11, 20
  – development in children 79–80, 107
  – evolution 9–10, 18, 21, 66, 106, 117,
    121, 125
Latin 4, 12, 88
lexical cohesion 47
lexical density 67, 76–77, 117
lexical item/word 76, 113
lexicogrammar 4, 6, 8, 32, 71, 109, 114,
  125
logical-semantic relations 55, 61, 64–65,
  91–92
  see also CONJUNCTION

Macro-New 249–252, 256–257, 264
Macro-Theme 249–252, 256–257
mathematics (language of) 12, 56–58,
  75, 81, 107
meronymy 126
  see also composition
metafunction 27–29, 90
  – and register 30
  experiential 119
  ideational 88, 90
  interpersonal 88, 90
  logical 91
  textual 88, 90, 119
metalanguages 111–112

metaredundancy 41–42, 46, 110
method of development 95, 244–247,
  252
mode 33, 34
MOOD 27

narrative 197–200
New 55, 60, 90, 92, 95, 98–102
  see also INFORMATION
Newton 6–8, 13, 57–62, 75, 81–82,
  88–90, 108
nominal group 7, 31, 55, 57, 63–64, 75,
  88–89, 101, 129–132, 154, 219–220,
  229
nominalization 15, 31, 39–40, 56, 64, 78,
  81, 98, 119, 127–129, 217–218

passive 58, 193–194
perspective 46–49
  see also dynamic, synoptic
phonology 32, 102
plane 24
point 247–249, 252
Priestley 62, 91
probabilities 109–110
PROCESS TYPE 27–28
processes,
  material 65, 89
  mental 65, 89
  relational 13, 42–43, 65, 88, 220,
    229–232
    attributive 153–154
    identifying 149–150, 222–224
  verbal 65
  see also transitivity
projection 38–39, 58, 65–66, 114
proof 61, 64, 77, 87, 90

rank 37–41
reality construction 9, 11, 15–21, 64, 82,
  92–93, 113, 118, 131–132
realization 17, 20, 41–45, 111, 114
redundancy 110, 117
register 15, 54, 86–87, 93, 124
reports 187–190, 206–208, 215–216,
  222–224
  classifying 187
  decomposing 187–188
rewording 66, 78–80, 83
  see also grammatical metaphor
rhythm 102

scientific knowledge 6, 15, 61–62, 67,
  81, 108

semantic,
 – discontinuity 82–84
 – space 87
 – variation 116
semogenesis 110–111, 125
social context 23–26
spoken language 21, 117–120
stratification 17, 29–32
syndromes 4, 8, 54–56, 66, 75, 86
synoptic (perspective) 48–49
system,
 – and instance 16–18, 109–111
 – and process 11, 18

taxonomy 126–127, 132, 137–143, 204,
  230
 folk /vernacular 6, 138–139
 scientific/technical 6, 64, 73–75, 138,
  141–142, 206
 see also classification, composition
technicality 3, 8, 56–59, 70, 118,
  143–145, 212–213, 225–226, 229
tenor 32–33, 35

TENSE 9, 109, 115
textual metafunction 29
THEME 241–242
Theme 55, 60, 63, 74, 90–92, 95–98,
  131, 211–212
Token 149–150, 222–224
transitivity 65, 88, 114, 129
 see also PROCESS TYPE, processes

universal (philosophical) language 5–6,
  108

Value 149–150, 222–224
variation,
 among languages 9
 within a language 54, 109
 see also dialect, register
verbal group 55, 61

Whorf 9, 17, 21, 113, 115
wording, *see* lexicogrammar
written language 11, 21, 107, 117–120